T0305964

Labor and Employment Issues for the Safety Professional

Occupational Safety and Health Guide Series

Series Editor

Thomas D. Schneid
Eastern Kentucky University
Richmond, Kentucky

Published Titles

Labor and Employment Issues for the Safety Professional

Thomas D. Schneid

CRC Press
Taylor & Francis Group
Boca Raton London New York

CRC Press is an imprint of the
Taylor & Francis Group, an **informa** business

CRC Press
Taylor & Francis Group
6000 Broken Sound Parkway NW, Suite 300
Boca Raton, FL 33487-2742

First issued in paperback 2017

CRC Press is an imprint of Taylor & Francis Group, an Informa business

No claim to original U.S. Government works

ISBN-13: 978-1-4398-2020-9 (hbk)
ISBN-13: 978-1-138-11768-6 (pbk)

Visit the Taylor & Francis Web site at
http://www.taylorandfrancis.com

and the CRC Press Web site at
http://www.crcpress.com

Contents

            Chapter Questions (True/False) 46
            Answers 46
            Endnotes 46

Chapter 8   Age Discrimination 49

            Chapter Questions (True/False) 52
            Answers 52
            Endnotes 52

Chapter 9   Wage and Hour Laws 55

            Chapter Questions (True/False) 58
            Answers 59
            Endnotes 59

Chapter 10 Federal Retirement and Welfare Laws 61

            Chapter Questions (True/False) 66
            Answers 66
            Endnotes 66

Chapter 11 Privacy Laws 67

            Chapter Questions (True/False) 70
            Answers 70
            Endnotes 70

Chapter 12 Family and Medical Leave Act and Safety 73

            Chapter Questions (True/False) 77
            Answers 77
            Endnotes 77

Chapter 13 Americans with Disabilities Act and Safety 79

            Chapter Questions 87
            Answers 88
            Endnotes 88

Chapter 14 OSHA Inspections and Defenses 91

            Chapter Questions 96
            Answers 96
            Endnotes 96

Foreword

Don't be afraid to think outside the book.

Dr. Schneid has put together a complete compendium of important topics, tools, and most importantly, practical thoughts.

The topic of labor and employment is an increasingly complex and changing field.

In your role as safety manager, human resources, security, or any or all of the above, this text is a tool that you will refer to daily. Its best use is to be on your desk, with pages underlined and marked up. You may be wearing many of these hats, all at the same time!

We don't operate in a vacuum in our profession, and neither should you.

Most labor situations are governed by a series of overarching federal rules and regulations. These rules are guidelines and are impacted by site agreements, interpretations, collective bargaining, and common sense.

This text is not a substitute for competent legal advice, and must not be used to avoid close working relationships with counsel. In fact, many problems may be averted by a careful review of the chapters that support the framework for how you work. And remember, no two situations are the same. Safety rules and regulations must be carefully interfaced into the entire environment as well. If you are unsure, seek advice from a competent professional, and don't forget that most states operate a free OSHA consultation program, as well as inside of federal OSHA. Ask advice; it's better to take the steps to protect your people with a team approach than choosing to ignore the situation. Afford your fellow workers the same opportunity to have a safe and healthy job experience. Would you do that same job, day in, day out, without protection or training? Education is the key, and training is but one component of that process. It evolves.

Human resources, safety, and security are a continuous process—no beginning, no end, just progress as the process moves forward!

Tom is always learning, and this book will help you to do just that.

As you scan the Table of Contents, don't be afraid of the myriad chapter headings.

These are but a few of the topics that come up in the daily life of a safety or HR manager. Supervisory staffs who have been promoted from the "line" to a more people-oriented position would do well to learn from this text, too.

The landscape is changing, almost on a daily basis, but this thorough text will help you navigate it in a more cohesive method.

A key point: Do not tread in these areas without a complete understanding of the subject.

New rules and interpretations change, cases make precedents, but the concepts that Tom has explicitly put before you are key to your understanding and survivability in the unique environment in safety, legal, and making a difference.

I have managed many facilities safety programs, as well as interfaced with human resources at all levels to do the best job we could. Tom and I helped get facilities in the OSHA VPP program together in the 1990s. Currently I work in homeland security, worldwide. Tom has remained a close friend and ally of mine for decades. He has been instrumental in my success, and we want to be part of yours.

Don't make the mistake of complacency and carelessness. Do your homework, and seek out competent professionals before the need arises in a crisis. Decisions made in haste can have significant impacts.

One key element of this wide-ranging text is that it will help you understand when you may need to reach out for support.

As I tell my students, the only poor question is one you don't ask.

Good luck as you use this book, and remember, highlight and turn the pages; use the book as but one of many tools in your tool box.

Tom will be glad you did, and so will you. You, your people, and your organization will benefit.

Be safe!

Michael J. Fagel, PhD, CEM*

---

* Mike is an instructor at Eastern Kentucky University, as well as at the University of Chicago and Northwestern University. He can be reached at mjfagel@aol.com.

Preface

Safety professionals do not work in a vacuum! The safety professional interacts with many other functions within the organization, including but not limited to production, human resources, and medical, with each function possessing specific laws and regulations that govern their actions and inactions. In order for a safety professional to function within the organizational structure, he or she should have a working knowledge of the laws and regulations that impact his or her area of responsibility as well as the laws and regulations that impact employees, managers, and the overall organizational structure.

In this text, we address many of the laws and regulations that impact the safety function, especially in the expanding areas of labor and employment laws. The primary reason for this emphasis is the fact that the safety function interacts daily with employees and management, and it is imperative that safety professionals possess a working knowledge of the impacts, requirements, and implications of their actions and inactions as related to these functions and laws. All too frequently very competent safety professionals get themselves "into the soup" simply because they are not aware of the new law, regulation, or interpretation, or they are not aware of their specific responsibilities or rights under the law. In this text, we cover a substantial number of the laws and regulations that can impact the safety function and hopefully provide at least an awareness level of the law or regulation for the safety professional to permit the "light bulb" to go off when he or she encounters situations where these laws or regulations are applicable.

The author reminds the reader that laws, especially in many of the areas that are addressed in this text, are constantly evolving and moving. Although the author has made every attempt to provide the most current information possible, safety professionals are advised to research the status of any specific issue and acquire competent legal assistance where necessary. The law is constantly changing, and it is imperative that safety professionals know the status of the law or regulation at the time of the issue. Remember, ignorance of the law is never a good defense; knowledge of the law can avoid many potential pitfalls!

The author hopes that this text will provide a working knowledge of these peripheral laws and regulations that can impact the safety function. Recognition and knowledge of these impacting laws and regulations can often help safety professionals avoid potential legal issues and possible legal liabilities for themselves as well as their organizations.

Disclaimer: The author has gone to great lengths to provide the most current information in a format that is most effective and efficient for safety professionals. However, the law changes virtually daily, and it is vital that the reader acquire competent legal counsel when addressing any of the topics or issues discussed in this text. The information contained in this text should in no way substitute for advice by your legal counsel.

Acknowledgments

Many apologies to my wife, Jani, and my girls, Shelby, Madison, and Kasi, for my absence at a number of basketball and softball games, as well as many social events, while I researched and drafted this text. And my thanks for your acceptance in permitting me the time on weekends and evenings to research and write this text without major drama. I know I will not get back the games I missed, but I will definitely be there for your future games.

Special thanks to my parents, Bob and Rosella, for branding the importance of education on their children and grandchildren. Without their continuous reinforcement and encouragement, your children and grandchildren would not have been able to reach the levels they have achieved in their lives.

Introduction

Whether you know the shape of a pebble or the structure of the solar system, the axiom remains the same: that it exists and that you know it.

—Ayn Rand

Safety professionals may be exposed to a number of different laws on a daily basis, depending on the circumstances, and often are not even cognizant of the law or the implications until after a potential violation has already occurred—and by this time it is often too late. This unconscious ignorance of the law can often result in legal actions against the company, the safety professional, or both. Knowledge of the applicable laws, or at least recognition of the potential liability, can often permit the safety professional to avoid the potential misstep or take proactive measures to minimize any possible damage. Conversely, failure to recognize the law and the legal requirement of the law can place the safety professional in a position where he or she is not aware of the violation and potential liability until it is often too late to avoid or minimize the potential damages. And in the long run, the results can be damaging not only in monetary terms for the company, but also in myriad ways to the safety professional and his or her career.

The purpose of this text is to educate and provide awareness to the safety professional with regard to a number of laws within labor, employment, and related areas that may impact him or her in the course of his or her daily activities. Safety professionals are not expected to become lawyers or even human resource professionals, but they must have a rudimentary knowledge of these laws to avoid "stepping on their tail." Simple recognition of the law and the basic requirements of the specific law can often permit safety professionals to avoid a potential pitfall or violation in their actions or inactions.

In this text, the author has attempted to identify and select several of the most pertinent laws that may impact a safety professional beyond the OSHA regulations; however, not all laws can be addressed within the confines of this text. It is the author's hope that this text will shed light on and open the eyes of safety professionals to the fact that we do not work in a vacuum, and that other laws do impact our daily safety-related activities. Awareness of the laws that can impact a safety professional is the first step in avoiding or minimizing a potential legal issue in the workplace.

The Author

Thomas D. Schneid is a tenured professor in the Department of Safety, Security and Emergency Management (formerly Loss Prevention and Safety) at Eastern Kentucky University and serves as the graduate program director for the online and on-campus master of science degree in safety, security and emergency management.

Tom has worked in the safety and human resource fields for over 30 years at various levels, including corporate safety director and industrial relations director. He has represented numerous corporations in OSHA- and labor-related litigations throughout the United States. Tom has earned a BS in education, MS and CAS in safety, as well as his juris doctor (JD in law) from West Virginia University and LLM (graduate law) from the University of San Diego. Tom is a member of the bar for the U.S. Supreme Court, 6th Circuit Court of Appeals, and a number of federal districts, as well as the Kentucky and West Virginia Bar.

Tom has authored numerous texts, including *Corporate Safety Compliance: Law, OSHA and Ethics* (2008), *Americans With Disabilities Act: A Compliance Guide* (1994), *ADA: A Manager's Guide* (1993), *Legal Liabilities for Safety and Loss Prevention Professionals* (2010), *Fire and Emergency Law Casebook* (1996), *Creative Safety Solutions* (1998), *Occupational Health Guide to Violence in the Workplace* (1999), *Legal Liabilities in Emergency Management* (2001), and *Fire Law* (1995). Tom has also coauthored several texts, including *Food Safety Law* (1997), *Legal Liabilities for Safety and Loss Prevention Professionals* (1997), *Physical Hazards in the Workplace* (2001), and *Disaster Management and Preparedness* (2000), as well as over 100 articles on safety and legal topics. He is currently completing work on a new text, *Labor and Employment Issues for Safety Professionals*, and is also coauthoring a text on legal issues in safety.

1 Labor, Management, and Safety

> If you want to be respected, be respectable.
> If you want to be liked, be likeable.
> If you want to be loved, be lovable.
> If you want to be employed, be employable.
>
> **—William J. H. Boetcker**

Learning objectives:

1. Acquire an understanding of the role safety professionals play within the organization.
2. Acquire an understanding of the laws related to unionization.
3. Acquire an understanding of the labor and management issues that directly or indirectly impact the safety function.

The workplace is a balance between management and labor, with the safety function often thrust into the middle. Generally, management is production driven, dollar cognizant, quality motivated, and salary paid. Again, generally labor consists of a number of individuals who perform specific work tasks, are responsible for individual job functions, and are paid by the hour. With these philosophically diverse positions, labor and management can conflict, creating adversarial situations that encompass the safety function. Safety professionals can be the link between management and labor, serving as the point to achieve the ultimate objective of creating and maintaining a safe work environment. In very broad terms, the safety professional is employed to minimize risks and save money for the company or organization. The safety professional is also employed to minimize risks and safeguard employees from injury or illness in the workplace. The safety professional must be able to work with and between all levels of management and labor, adapting to the different motivational factors and conflicting issues.

In today's American workplace, a safety professional is more likely to be working in a nonunion environment than in a unionized private sector environment. Although this was substantially different a half century earlier, union membership in the private sector has fallen to under 9% of the private sector workplace, the lowest membership since 1932.[1] Although there are many different unions representing different occupations, the vast majority of unions are aligned with one of the two larger national and international organizations; namely, the American Federation of Labor–Congress of Industrial Organizations (AFL-CIO) and the Change to Win Federation.[2] The AFL-

CIO and Change to Win Federation are both active in the political arena, advocating for policies and legislation favoring their membership.

The primary law that regulates the private sector in the area of unionization is the National Labor Relations Act.[3] This law, as well as other related laws governing this area, is overseen by the National Labor Relations Board (NLRB), an independent federal agency.[4] The National Labor Relations Board has established rules and regulations governing the election process, unfair labor practices, and other areas of the collective bargaining process and the relationship between private sector employers and the union.

Safety professionals should be aware that there are very specific rules that must be followed in the event that a union attempts to organize an operation, and they must be very cognizant of the rules as well as their actions or inactions, given the fact that safety is often one of the major issues in which employees may seek union representation. The election process, which includes laboratory conditions, is proscribed and overseen by the National Labor Relations Board. After the election, if the union has won a majority of the votes of the bargaining unit and is certified in the workplace, it is the duty of the company or organization to bargain in good faith with the union over wages, hours, and conditions of employment, which includes safety. If an agreement is reached, these negotiated terms and conditions are codified in writing and become a legally binding contract. Many contracts call for disputes over the contract to be resolved through a grievance process or arbitration.

For safety professionals, the development and implementation of effective safety and health programs can be significantly different in a nonunion environment than in a unionized workplace. As discussed later in this text, safety and health issues are considered part of the collective bargaining process in unionized operations; thus mandatory bargaining is often required before implementing safety and health programs, policies, and procedures in the workplace. In comparison, safety professionals working in nonunion working environments usually do not possess the duty to bargain with a third party prior to the development and implementation of safety and health programs, policies, and procedures.

Although the goals and objectives of safeguarding employees in the workplace may be the same, the methods of achieving these goals may be significantly different for safety professionals, depending on whether the operations are unionized or without union representation.

The vast majority of safety professionals are members of the management team hired specifically to address an issue or issues within the management structure. Although many larger companies or organizations have previously integrated safety into their management structure because it made good business sense, smaller companies or organizations do not realize the need for a safety professional until after a major risk or loss surfaces. These risks or losses can include, but are not limited to, Occupational Safety and Health Administration (OSHA) violations and penalties, workers' compensation losses, substantial property or casualty losses, insurance increases, or related issues. The acquisition of a safety professional is often part of the overall mitigation efforts after an incident or loss. If a company or organization possessed no losses and maintained a minimal risk level, the organization may not have a need for the costs related to employing a safety professional. In essence, companies and organizations employ safety professionals to manage their risks and minimize their exposure to potential losses. Management of these risks entails, in part, the creation and maintenance of a safe and healthful work environment that is in compliance with all governmental regulations and requirements.

The safety professional, as a member of the management team, often serves in a consulting role at various corporate levels within the management hierarchy. Safety professionals often wear many official as well as unofficial hats within an organization. Officially, safety professionals are often responsible for the safety and health of employees, compliance with applicable governmental regulations, workers' compensation, security, environmental issues, and myriad other combinations. Unofficially, the safety professional can serve in a broad spectrum of roles, including as the listener of all complaints, the marriage counselor for upset employees, the financial researcher for the supervisor, and the company softball team manager. These official and unofficial roles of the safety professional can place him or her in a possible conflict situation between the various demands and obligations of each of the roles he or she plays in the workplace.

The safety professional's official job description may be the same in a unionized environment as in a nonunion environment; however, the activities and methods utilized to achieve the safety and health goals and objectives can be very different. In a unionized environment, safety and health are a mandatory subject within the scope of collective bargaining, and thus all aspects of the safety and health effort would be part of a negotiated process between the safety professional, as an agent for the company or organization, and the union, as a third-party collective bargaining representative of the employees.

Conversely, in a company or organization where the employees are not represented by a union, the safety professional may be more of an advocate for employees with management in the area of safety, since there is no third-party representation and negotiation process. The safety and health efforts of safety professionals can be very effective in a nonunion environment as well as in a unionized environment; however, the route through which to achieve the goals and objectives can be very different.

The safety and health of employees are of vital importance, whether in a unionized environment or nonunion environment. Although the prospective and methodologies may be different, unions and most management teams identify safety as one of, if not the highest, priorities in the workplace. Safety professionals, as members of the management team, play a key role in developing, implementing, and enforcing safety programs, procedures, and policies to safeguard employees in the workplace. The role of the safety professional as a member of the management team, an advocate for employee safety and health, an advocate for management in negotiations with the union, a manager of specific other functions (including security, workers' compensation, etc.), a governmental agency liaison, and other activities can create internal and external conflict for the individual safety professional.

Safety professionals working in the private sector are unique and are a breed in and of themselves. Safety professionals are often the only management team member responsible for the entire safety and health function, and thus often work alone or with a very small staff. Often the management team possesses limited knowledge of the functions within the safety and health area and relies on the safety professional as the "on-site expert" in any issues or situations involving safety and health.

Safety professionals are often required to define their own role within the management hierarchy and are frequently provided little oversight or assistance at the operational level. Safety to many management teams is a nebulous concept that creates difficulties and is the responsibility of the safety professional. One of the most difficult challenges for many safety professionals is to inject change within the management team structure to accept responsibility for the safety and health of their employees. Safety professionals should provide the knowledge, the tools, and the methods to shift this level of responsibility to the management team members and develop a methodology through which to hold them accountable for their performance in the area of safety and health.

Safety professionals in the private sector are almost universally responsible for achieving and maintaining compliance in the workplace with the standards and regulations established by OSHA. Although the OSHA standards are developed and promulgated to address workplace hazards that could result in injury or illness, the programs developed by safety professionals through which to achieve and maintain compliance can often be a source of conflict within unionized as well as nonunion operations. The OSHA General Industry Standards establish a base level of compliance below which no safety professional should permit his or her program to fall. Most safety professionals strive to achieve a level far above the basic compliance levels with their safety and health programs.

The OSHA standards do not specify how a safety professional must achieve compliance, but only that compliance must be achieved and maintained. The OSHA General Industry Standards are basically written to encompass all industries; thus the term *general industry* is utilized. The OSHA standards establish that the safety and health programs must reach and maintain a designated level; however, the methods of achieving and maintaining the designated level may vary widely between a unionized and nonunion operation. In essence, OSHA established that the safety and health efforts must reach and maintain a level of Z. A safety professional in a nonunion operation may use steps A, M, N, P, and Q to reach and maintain the level of Z, while a unionized operation may use steps A, B, C, L, M, N, R, S, and T to achieve and maintain the level of Z. The safety professionals both achieved and maintained the required level; however, the method through which the required level was attained varied with the different issues and requirements of the unionized and nonunion operations.

One common responsibility of safety professionals that can create conflict in unionized as well as nonunion operations is that of workers' compensation administration. In theory, the management of the safety and health function in most operations is a proactive activity, e.g., prevention of injuries and illnesses. Workers' compensation administration is a reactive activity resulting from the injury or illness already having taken place and the monetary compensation for the injury or illness. Although the safety function and the workers' compensation are linear in nature, resulting from a cause and effect, these functions are substantially different and can create opposition and conflict in both unionized and nonunion operations when both functions are managed by a safety professional.

Many companies and organizations have identified this inherent conflict and separated the workers' compensation and safety functions. Although workers' compensation is an individual state program with individual state laws, regulations, and monetary payments, the workers' compensation function can be an area of conflict as well as litigation in unionized as well as nonunion operations. In unionized operations, workers' compensation can also be a source of grievance by employees. The inherent conflict resulting from workers' compensation issues between the company and the employee or union can create barriers that can detrimentally affect the proactive efforts within the safety function.

Safety professionals must remember that they do not work in a vacuum. The internal structure of the organization, whether the employees are represented by a union or not, will dictate the methodologies utilized to achieve the safety objectives. Although most safety professionals possess responsibilities for achieving and maintaining compliance with the OSHA regulations, there are myriad other laws and regulations that impact the safety function directly or indirectly on a daily basis. And there are external events, such as economic downturns, which are outside of the control of the safety professional, that can impact the achievement of the safety objectives. In essence, a safety professional is the ultimate "juggler," providing maximum flexibility and instantaneous modification in order to maintain "all of the balls in the air." Although developing, maintaining, and managing safety within a unionized environment is significantly different than in a nonunionized environment, safety professionals, within the lines of management and the union, have proven that exceptional safety and health programs can be developed and maintained that achieve the goals and objectives of management, employees and their representatives, shareholders, and all other affected parties.

The world is changing. Although most safety professionals in the private sector work in an environment where a union is not present today, this was not the case less than 50 years ago. Safety professionals should be aware that this environment may change dramatically with the passage of several proposed laws, or even a subtle change in the way in which companies manage their nonunion employees. If history has taught us anything, it is that change is inevitable. The pendulum has swung toward the position that employees do not need union representation in the workplace because they are receiving a fair wage, good benefits, and are being treated fairly by management. The pendulum can swiftly swing in the opposite direction if any of these conditions change in any manner or the laws change to alter the current conditions. Safety professionals should be prepared to make the appropriate changes to continue to create and maintain a safe and healthful workplace for all of their employees.

CHAPTER QUESTIONS (TRUE/FALSE)

1. Labor organizations represent under 9% of the employees in the American private sector.
2. The AFL-CIO and Change to Win are the largest national labor organizations in the United States.
3. Safety is a required issue in collective bargaining negotiations.

4. The NLRB is the agency enforcing the FMLA.
5. The NLRB governs the "laboratory conditions" during a campaign and election.

ANSWERS

1. True
2. True
3. True
4. False
5. True

ENDNOTES

1. Bureau of Labor Statistics (January 28, 2008).
2. Change to Win Federation split from the AFL-CIO in 2005.
3. The National Labor Relations Act is also known as the Wagner Act.
4. Unionization in the public sector is regulated by both federal and state laws.

2 At-Will Employment and Exceptions

The employer generally gets the employees he deserves.

—Sir Walter Bilbey

Are soft-hearted people handicapped in business? You have heard a businessman say of someone else, "He's all right, but he's too soft-hearted...." To be soft-hearted may be handicapping, in a sense. But on the whole, a soft heart is to be preferred to a hard heart. Hard-hearted, severe, dominating giants sometimes manage to get further and to amass more money. But they get less genuine joy out of life.... It is the hard-boiled employer, not the soft-hearted species, that incites most of our strikes and does most to endanger the harmonious progress of democracy.

—B. C. Forbes

Learning objectives:

1. Acquire an understanding of the concept of an employee in the American workplace.
2. Acquire an understanding of the at-will doctrine and exceptions.
3. Acquire an understanding of the employer-employee relationship requirements for OSHA.

Who is an employee in the eyes of the law? What rights does an employee have in the American workplace? Are the rights for an employee represented by a union different than those for an employee working in a nonunion operation? How is an employee identified in the workplace? Who is an employee in the eyes of OSHA? Identifying the appropriate employment status of an individual in the workplace often creates conflicts for safety professionals that can lead to grievances, litigation, and additional internal and external difficulties.

The workplace of today can be significantly different than the workplace of our fathers and grandfathers. As identified in the *Report and Recommendations of the Commission on the Future of Worker-Management Relations*,[1] "As employers seek new ways to make the employment relationship more flexible, they increasingly rely on a variety of arrangements popularly known as 'contingent work'."

The use of independent contractors and part-time, temporary, seasonal, and leased workers has expanded tremendously in recent years. On the positive side, contingent employment relationships are, in many respects, a sensible response to today's competitive global marketplace; various forms of contingent work can offer benefits to both management and workers. Contingent arrangements allow some firms to maximize workforce flexibility in the face of seasonal or cyclical forces and demands of modern methods, such as just-in-time production. This same flexibility helps some workers who must balance the demands of family and work as the number of dual-earner and single-family households rises. On the negative side, contingent arrangements may be introduced simply to reduce the amount of compensation paid by the firm for the same amount and value of work. This is particularly true because contingent workers are drawn disproportionately from the vulnerable sectors of the workforce. They often receive less pay and benefits than traditional full-time or "permanent" workers, and they are less likely to benefit from the protections of labor and employment laws. A large percentage of workers who hold part-time or temporary positions do so involuntarily.[2]

It is important that safety professionals be able to recognize the type of employment relationship that has been created between a company or organization and the individual working in the workplace.

Although in most organizations the individual performing the work activity is an employee, there are significant differences between an at-will employee, a union employee working under a collective bargaining agreement, a worker contracted through a temporary employment agency, and an individual performing work under a contract. Safety professionals should be able to function within the bounds of the significantly different requirements for each of these types of individuals working within the operation, as well as ensuring compliance with the law, no matter which type of employment relationship is utilized in the workplace. The difficulty often encountered by safety professionals is that the employment relationship type is not readily apparent by the work activity, clothing, safety equipment, or other visibly recognized method. The visual signs that safety professionals often utilize within many operations, such as the color of the hard hat or the color of the uniform, can be an unreliable method, depending on the workplace. At-will employees, temporary contract employees, and subcontractors may be working side by side wearing the same safety equipment, uniform, and personal protective equipment (PPE).

Safety professionals often must identify the individual worker's employment status from the human resources department or other department, in order to identify the unique employer-employee relationship of individuals working in the workplace. Safety professionals should consider the below listed basic questions, which can be utilized to assist in identifying the type of employment relationship that is present in most situations. Safety professionals should identify the following:

1. Is this individual an employee of your company? If he or she is *not* an employee, is there a contract between the contractor and your company?
2. Is the employee an at-will employee?
3. If the individual is an employee of your company, what employment type is the employee?
4. Is the employee covered under a collective bargaining (union) contract?
5. Is there an employee handbook?
6. Is the individual a temporary employee?
7. Is the employee a probationary employee?

8. Is the individual a contractor? Is the individual employed by another company but performing work within your operation?
9. Is the individual a visitor to the company or organization?

Safety professionals should be aware that each of the above creates different requirements when addressing safety in the workplace. Each of the identified types of employment relationships provide specific managerial challenges and requirements when attempting to develop, implement, and manage the safety function within the operations. Each type of potential employment or contract relationship creates different requirements the safety professional must manage in order to accomplish his or her safety-related activities and goals. Acquiring a working knowledge of the different employment relationships in the workplace will permit the safety professional to identify the procedural requirements necessary, as well as keep the safety professional from creating conflict and difficulties when attempting to develop, implement, manage, or enforce safety policies, programs, and procedures within the operations.

In today's American workplace, the most common type of employment relationship is that of an at-will employee. An at-will employee does not belong to a union (and thus is not included within the collective bargaining agreement), is not working under contract for a temporary employment agency, does not possess a written contract of employment with the company or organization (usually reserved for executives or consultants), and usually works within the proscribed company policies or the policies established within the company's written employee handbook. In general, every state in the union recognizes at-will employment in varying forms, and this type of employee can be terminated for good cause, bad cause, or no cause at all, unless the employee is terminated for one or more of the provided exceptions created by law or provisions specified within the company's employee handbook.

Historically, the at-will employment doctrine was derived from the master-servant relationship in Europe, primarily England, in the Middle Ages. Since the 1800s, the at-will employment doctrine emerged as the common-law rule in the United States, with exceptions for termination of the employment relationship being created by the courts. In at-will employment any employee who does not possess a contractual agreement to the contrary, or protection under federal or state statutes, is considered to be employed at the will of the employer and may be terminated at any time, without notice and with or without just cause.[3] This doctrine was codified in several early court decisions and has become established within the American judicial process as well as in the American workplace.[4] Over the years, however, the stringency and inequity of the at-will doctrine has been eroded through federal and state laws prohibiting discrimination or retaliation, including, but not limited to, the Labor-Management Relations Act[5] and Title VII of the Civil Rights Act of 1964,[6] as well as the federal and state courts creating judicial exceptions to this general rule. Some of the current court-recognized exceptions available in many states today include the public policy exception (e.g., refusal to commit an unlawful act), implied convenience of good faith and fair dealing, good cause requirements for termination, implied conditions in personnel manuals and handbooks, tortious discharge exception (e.g., outrageous circumstances), and promissory estoppels (e.g., reneging on a promise).[7] Safety professionals should be aware that although there are some common exceptions, each state has recognized specific exceptions for its jurisdiction. A state-by-state listing of the identifiable exceptions are as follows:

Alabama
- Independent considerations exception
- Implied contract exception
- Intentional infliction of emotional distress exception

Alaska
- Covenant of good faith
- Permanent employment exception

Arizona
- Public policy exception
- Implied contract exception

Arkansas
- Public policy exception
- Intentional infliction of emotional distress exception

California
- Implied contract exception
- Covenant of good faith and fair dealing exception
- Independent consideration exception
- Public policy exception

Colorado
- Has not adopted exceptions to the historic employment at-will doctrine (1986)

Connecticut
- Violation of public policy
- Covenant of good faith and fair dealing
- Independent consideration exception

Delaware
- Courts have refused to judicially erode the employment at-will principle (1986)

District of Columbia
- Courts have been reluctant to recognize judicial exceptions to the employment at-will doctrine (1986)

Florida
- Public policy exception
- Independent consideration exception

Georgia
- Courts have refused to recognize deviations from the employment at-will doctrine (1986)

Hawaii
- Public policy exception
- Promissory estoppel exception

Idaho
- Implied contract exception
- Fraud

Illinois
- Public policy exception
- Independent consideration exception
- Intentional infliction of emotional distress exception

Indiana
- Public policy exception
- Promissory estoppel exception

Iowa
- Covenant of good faith and fair dealing exception
- Independent considerations exception

Kansas
- Public policy exception
- Promissory estoppels exception
- Implied contract exception

Kentucky
- Public policy exception
- Implied contract exception
- Presumption of term exception

Louisiana
- Courts have refused to recognize some exceptions and have specifically rejected public policy and implied contracts exceptions (1986)

Maine
- Courts have been reluctant to subscribe to some developing exceptions; public policy exception has been rejected (1986)

Maryland
- Public policy exception
- Intentional infliction of emotional distress exception
- Implied contract exception

Massachusetts
- Public policy exception
- Implied contract exception
- Covenant of good faith and fair dealing exception
- Intentional infliction of emotional distress exception

Michigan
- Public policy exception
- Negligent discharge exception
- Intentional infliction of emotional distress exception
- Implied contract exception

Minnesota
- Promissory estoppels exception
- Implied contract exception
- Independent considerations exceptions
- Intentional infliction of emotional distress exception
- Covenant of good faith and fair dealing exception

Mississippi
- Implied contract exception

- Independent consideration exception
- Intentional tort exception

Missouri
- Public policy exception
- Intentional infliction of emotional distress exception
- Implied contract exception

Montana
- Covenant of good faith and fair dealing exception
- Public policy exception

Nebraska
- Courts have given little deference to some judicial exceptions, but have implied they may be less sympathetic to strict adherence to the at-will doctrine in certain cases in the future (1986)

Nevada
- Public policy exception
- Implied contract exception

New Hampshire
- Covenant of good faith and fair dealing exception
- Public policy exception
- Independent consideration exception

New Jersey
- Public policy exception
- Implied contract exception

New Mexico
- Public policy exception
- Implied contract exception

New York
- Implied contract exception
- Promissory estoppels exception

North Carolina
- Implied contract exception
- Public policy exception

North Dakota
- Courts have been reluctant to recognize some judicial exceptions to the at-will doctrine (1986)

Ohio
- Implied contract doctrine
- Promissory estoppel exception
- Public policy exception

Oklahoma
- Implied contract exception

Oregon
- Implied contract exception
- Public policy exception
- Intentional infliction of emotional distress exception

Pennsylvania
- Public policy exception
- Independent consideration exception
- Intentional infliction of emotional distress exception

Rhode Island
- Courts have consistently refused to recognize some of the developing exceptions to the at-will doctrine (1986)

South Carolina
- Courts continue to apply the at-will rule in discharge matters (1986)

South Dakota
- Implied contract exception
- Presumption of term exception

Tennessee
- Courts generally adhere to the at-will principle
- Public policy exception
- Implied contract exception

Texas
- Public policy exception
- Implied contract exception

Utah
- Courts have been the most reluctant to recognize exceptions to the at-will rule (1986)

Vermont
- Courts have not recognized any of the developing exceptions to the at-will principle (1986)

Virginia
- Public policy exception
- Implied contract exception
- Presumption of term exception

Washington
- Public policy exception
- Implied contract exception
- Intentional infliction of emotional distress exception

West Virginia
- Public policy exception
- Implied contract exception
- Intentional infliction of emotional distress exception

Wisconsin
- Implied contract exception
- Public policy exception

Wyoming
- Courts have strictly adhered to the at-will doctrine (1986)

The vast majority of safety professionals working in the private sector American workplace are employed as at-will exempt (salary) managerial employees. Thus, safety professionals should be aware that their employment with the company or organization can be terminated at any time. Some safety professionals have opted to negotiate individual employment contracts with their employers identifying specific contractual requirements for termination in order to avoid the categorization as an at-will employee in situations involving termination. Safety professionals should be aware that many employers will not negotiate this type of employment contract in order to maintain a greater latitude in terminating employees.

Given the substantially large percentage of nonunion companies and operations in the United States, the majority of employees work as at-will employees, often with employment requirements set forth in an employee handbook or posted policies. Safety professionals should be aware that most employee handbooks or company policies usually contain a provision identifying the at-will status, thus maintaining the employer's ability to terminate any employee at any time for any lawful reason. It is important that safety professionals identify this employment status and become fully knowledgeable with the employee handbook or company policies, especially in the provisions addressing safety, security, and disciplinary actions.

Safety professionals should also be aware that the employment status also has a direct bearing on the safety function as identified in the Occupational Safety and Health Act of 1970. In the OSH Act, the creation of an employer-employee relationship is essential in order for the employer to potentially be found liable for violations of the OSH Act by OSHA. The primary question that often arises is whether a company or organization can be found liable for alleged violations under the OSH Act when its own employees are exposed to a hazard created by another employer or when the employer responsible for creating the hazard does not affect any of its own employees. Safety professionals have often utilized the lack of an employer-employee relationship as one of the defenses against an issued violation.

The Occupational Safety and Health Review Commission (OSHRC) initially addressed the employer-employee relationship shortly after the OSH Act was enacted in the case titled *Gilles v. Cotting, Inc.*[8] In this case, two employees of the primary contractor were killed in an accident created by and resulting from the actions of a subcontractor. OSHA issued one citation to the primary contractor and one citation to the subcontractor. OSHA subsequently vacated the citation to the primary contractor, finding the citation was improper and would impose "liability outside the employment relationship."[9] Over the years, the OSHRC, as well as the various federal Circuit Courts of Appeal, has addressed this issue primarily in the construction area, with focus on creation of the hazard and control of the hazard.[10] In recent decisions and with the inception of the multiemployer worksite rule, the scope of the duty to safeguard employees at the worksite has been substantially expanded.[11]

Safety professionals should be aware that under the OSHA multiemployer worksite policy, more than one employer can be cited for alleged violations on a worksite.[12] Specifically, the OSHA compliance inspector is instructed to conduct a two-step analysis to determine whether more than one employer can be cited for an alleged violation. The two-step process includes:

> Step One. The first step is to determine whether the employer is a creating, exposing, correcting, or controlling employer. The definitions in paragraphs (B)–(E) below explain and give examples of each. Remember that

an employer may have multiple roles (see paragraph H). Once you determine the role of the employer, go to Step Two to determine if a citation is appropriate (NOTE: only exposing employers can be cited for General Duty Clause violations).

Step Two. If the employer falls into one of these categories, it has obligations with respect to OSHA requirements. Step Two is to determine if the employer's actions were sufficient to meet those obligations. The extent of the actions required of employers varies based on which category applies. Note that the extent of the measures that a controlling employer must take to satisfy its duty to exercise reasonable care to prevent and detect violations is less than what is required of an employer with respect to protecting its own employees.[13]

Of particular importance to safety professionals is the determination whether your company or organization created the alleged hazard, exposed your employees to the alleged hazard, is responsible for correcting the alleged hazard, or controls the workplace. This is especially applicable for safety professionals with worksites where contract employees, subcontractors, temporary workers, loaned workers, or other workers who may not qualify as an employee are working at the company or organization's worksite.

For employees who have formed or joined a recognized union, safety professionals should be aware that the employment status of these employees can be significantly different than that of an at-will employee. An employee who is a member of a union and encompassed within a collective bargaining agreement is governed by the negotiated terms and conditions set forth in that agreement (also known as the union contract). Safety professionals should be aware that the at-will doctrine is no longer applicable, and the primary basis of this employment relationship lies in the area of contracts.

In general terms, safety professionals should be aware that the union is the representative of the employees identified within the specific collective bargaining unit. After recognition of the union, the company or organization possesses a duty to bargain in good faith for wages, hours, and conditions within the collective bargaining unit. Safety in the workplace is considered a condition of employment, and thus safety programs, policies, and other aspects of the safety function are included within the negotiations. Of particular importance to safety professionals are the negotiated terms with regards to disciplinary action for safety actions or inactions by employees. The terms and conditions of any disciplinary action would be a negotiable issue, and thus established by the collective bargaining agreement.

For temporary or contract workers, safety professionals should be aware that the worker's status is often determined by the terms and conditions agreed upon in a written contract. Although the worker may be working within your operations, the worker may actually be an employee of another company or entity provided to your company to perform specific tasks under a written contract. Although there may be many different variations of the contract terms and conditions, safety professionals should be cognizant to ensure that contract workers are properly trained, properly equipped, and the appropriate workers' compensation or insurance coverage is addressed within the written contract.

For most safety professionals, the employment status of the workers in their operation is often inconsequential since the safety and health of *all* workers is of the highest priority. However, safety professionals may find that the type of employment status is vitally important, especially in the area of discipline for safety violations, compliance program develop and implementation, and compliance program management. Focusing initially on the area of discipline, in many nonunion companies or organizations where workers are at-will employees, the company or organization has established progressive disciplinary systems wherein employees could receive disciplinary action, up to and including discharge for violation of established safety rules and policies. This progressive disciplinary system often includes several levels of possible disciplinary action, including verbal warnings, written warnings, suspension without pay, and involuntary termination. Companies and organizations often provide the disciplinary policy to employees when hired, post it for continuous review, and place it in the employee handbook. The level of disciplinary action for the violation of the safety rule or policy is often at the discretion of management.

For employees working under a collective bargaining (union) contract, the disciplinary policy and procedures are a negotiable issue between the company and the union that is usually codified in the collective bargaining contract. The negotiated disciplinary system can vary widely; however, it usually possesses grievance and arbitration provisions at the higher levels within the disciplinary structure. Given the negotiated nature of the disciplinary system in unionized operations, it is important for safety professionals to become well versed in the collective bargaining agreement terms and conditions, especially in the area of disciplinary action for safety rule violations.

Many companies and organizations contract with temporary employment agencies or other entities to acquire workers to perform specific tasks within the workplace. Safety professionals should be aware that there is usually no negotiations with the temporary employment agency or related entities with regards to safety programs, policies, or procedures; however, the area of disciplinary action for violation of safety rules can create issues for the safety professional. Most temporary or contract workers are employees of the temporary employment agency or other related entity performing work at your worksite. The negotiated contract is between your company or organization and the temporary employment agency or related entity wherein your company pays a fee and the temporary employment agency provides the required manpower for a specified job and a specified time period. In most cases, the individual employed by the temporary employment agency is an at-will employee of the temporary employment agency, and thus the temporary employment agency provides workers' compensation coverage and provides payment for services to the individual and related activities. Many contracts require the employee of the temporary employment agency to adhere to all safety policies and procedures within the company or organization's workplace. However, when the safety professional observes a temporary employment agency employee violate a safety rule or policy, what is the procedure to initiate disciplinary action against the individual when he or she is not an employee? In many cases, the safety professional will be required to contact the temporary employment agency and require the agency to initiate the disciplinary action as required under the terms and conditions of the contract.

What often appears to be a crystal-clear situation for the safety professional can often be a minefield in disguise. It is important for safety professionals to become knowledgeable regarding the employment status of the individuals working in their operations in order to avoid unnecessary grievances, contractual entanglements, and the potential of litigation, especially in the area of disciplinary action for safety violations. Prudent safety professionals may want to discuss this issue with their human resources department or legal counsel in order to ensure that they have a firm grasp of the various employment statuses of individuals in their workplace and the proper procedures to follow in all phases of their program development, implementation, and enforcement. The private sector American workplace still predominantly functions under the at-will employment doctrine; however, unionization and contractual relationships can alter this predominant view. With employment status, there is little room for a safety professional to guess as to the status. Safety professionals should be fully knowledgeable of each and every worker's employment status and utilize the proper procedures to successfully implement and enforce all phases of the safety and health efforts.

CHAPTER QUESTIONS (TRUE/FALSE)

1. An independent contractor is considered an employee.
2. An at-will employee is represented by a labor organization.
3. Every state recognizes at least one exception to the at-will doctrine.
4. The OSH Act required an employer-employee relationship.
5. At-will employees cannot be terminated for any reason.

ANSWERS

1. False
2. False
3. True
4. True
5. False

ENDNOTES

1. U.S. Department of Labor and Commerce, The Dunlop Commission, *Report and Recommendations of the Commission on the Future of Worker-Management Relations*, December 1994; also see Estreicher, W., and Schwab, S., *Foundations of Labor and Employment Law*, Foundation Press, 2000.
2. Ibid.
3. Wood, H. G., *Law of Master and Servant*, 2nd ed., Section 134, J.D. Parsons, Jr., Albany, NY, 1877, pp. 272–273.
4. See Payne v. Western & Atlantic Railroad, 81 Tenn. 507, 519–520 (1884): "All employers may dismiss their employees at will, be they many or few, for good cause, for no cause, or even for cause morally wrong, without being thereby guilty of legal wrong."

5. 29 USCA Section 151, et. seq.
6. 42 USCA Sections 2000e-2 and 2000e-3.
7. Jackson, G. E., *Labor and Employment Law Desk Book*, Prentice-Hall, 1986.
8. 1 OSHC 1388, revised sub nom., *Brennan v. Gilles & Cotting, Inc.*, 504 F.2d 1255 (4th cir.), on remand, 3 OSHC 2002 (1976).
9. Ibid.
10. See *Brennen v. Gilles & Cotting, Inc.*, N. 69, supra; *Brennan v. OSHRC*, 513 F.2d. 1032 (2nd Cir. 1975), *Marshall v. Knutson Construction Co.*, 566 F.2d 596 (8th Cir. 1977).
11. See Summit Contractors, OSHRC Docket 03-1622 (2009); Munro Waterproofing, Inc., 5 OSHC 1522 (1977); Hopkins Erection Co., 5 OSHC 1034 (1977).
12. CPL 02-00-124, December 10, 1999.
13. Ibid.

3 Labor Law and Safety

The lottery of honest labor, drawn by time, is the only one whose prizes
are worth taking up and carrying home.

—**Theodore W. Parker**

The dignity of labor depends not on what you do, but how you do it.

—**Edwin Osgood Grover**

Learning objectives:

1. Acquire an understanding of the various labor laws in the United States.
2. Acquire an understanding of the different types of union security arrangements.
3. Acquire an understanding of the NLRB and its role in the election process.
4. Acquire an understanding of unfair labor practices.

The forerunner of today's unions started with the guilds and skilled craftpersons in Europe in the Middle Ages and emerged in the United States in the eighteenth century. These early trade craft groups, who banded together for the purposes of negotiating with an employer over such issues as wages and assistance for injury or death, were often found to be illegal. With the working conditions changing with the emerging industrialization of the United States in the late nineteenth and early twentieth centuries, in conjunction with such events as the Great Depression, World War I, and the New Deal, an environment and conditions were created that were favorable for the growth of unions and the creation of workers' rights in the industrial workplace.

The current labor relations law was created by the laws enacted during this turbulent period of time, namely, the Norris-LaGuardia Act of 1932, the National Labor Relations Act (NLRA; also known as the Wagner Act of 1935), the Labor-Management Relations Act of 1947 (also known as the Taft-Hartley Act), and the Labor-Management Reporting and Disclosure Act of 1959 (also known as the Landrum-Griffin Act). Although there were a number of earlier laws, including the Sherman Antitrust Act of 1890, the Clayton Act of 1914, the Railroad Labor Act of 1926, and the National Industrial Recovery Act of 1933, that laid the groundwork, the framework for today's labor-management relations and collective bargaining process was established by the combination of the National Labor Relations Act, the Labor-Management Relations Act, and the Labor-Management Reporting and Disclosure Act. Although there are myriad other laws that impact the labor-management relations area in various ways today, including such laws as the Occupational Safety and Health Act and the Civil Rights Act, the basic foundation and framework of labor law in the United States remains in the Wagner Act, Taft-Hartley Act, and Landrum-Griffin Act.

Safety professionals usually are not involved directly in the relationship between labor and management; however, a base-level knowledge of the important aspects that do impact the safety function are essential. A labor organization, also commonly called a union, can be the representative of the employees who choose or are required to become members for the purposes of collective bargaining or negotiating with the company or organization over mandatory subjects, including wages, hours, and conditions of employment. In addition to the federal labor laws, many states have passed right-to-work laws that outlaw compulsory union membership and the required dues paid by employees.

Following the specific labor laws identified above, as well as the procedures established by the National Labor Relations Board (NLRB), employees can form or join a labor organization. If elected, the labor organization becomes the employee's representative. The labor organization and the company or organization are required to negotiate to impasse over the wages, hours, and conditions of employment (which include safety and health). These negotiations often lead to a collective bargaining agreement, or union contract, which contractually establishes the rules of the workplace. Members of the labor organization often pay a membership fee to join and a periodic payment, usually referred to as dues, to maintain membership.

Safety professionals should be aware that there are different types of union security arrangements that affect the payment of dues by employees. First, safety professionals should be aware of the Taft-Hartley Act, which made "closed shops," or workplaces requiring membership in the union prior to employment, illegal. However, several other types of union security arrangements are permitted to be included within a collective bargaining agreement. In a union shop, the company or organization is free to hire whomever it chooses; however, employees hired for positions covered by the collective bargaining agreement are required to join the labor organization after a specified probationary period. The union shop is the strongest relationship between labor and management currently permitted under law. An agency shop is a union security arrangement where nonunion employees are required to pay the union monies equal to the union dues or fees as a condition of continuing employment with the company. The purpose of this payment of union dues despite not being a union member is to compensate the union for its collective bargaining work and the company's desire to make union membership voluntary. The most common and least desirable form of union security arrangements is an open shop. In an open shop, membership in the union is voluntary, and employees choosing not to belong to the union are not required to pay dues.

In very general terms, safety professionals should be aware of the elongated process governed by the National Labor Relations Board, through which employees can vote to elect or not to elect a labor organization to represent them in the workplace. This general process includes the following steps:

1. Employees contact the labor organization.
2. The labor organization files a petition with the National Labor Relations Board.
3. The bargaining unit is determined by the NLRB.

4. Union authorization cards are signed by employees. (Note: If over 50%, the union can ask for voluntary recognition of the union without an election.)
5. The company is notified. An *Excelsior* list of employees is usually requested.
6. The NLRB establishes "laboratory conditions" in the workplace.
7. There are very specific rules during the preelection period.
8. Picketing, solicitation, boycotts, and campaigning take place.
9. Unfair labor practice charges are filed by the company and union.
10. The NLRB holds the election.
11. If the labor organization wins the election, collective bargaining negotiations begin.
12. If the company wins, the union is barred for a period of time.

The foundational element of which safety professionals should be aware in the Labor-Management Relations Act and the Wagner Act is the employee rights section (Section 7 of the Labor-Management Relations Act), which provide protection for employees to form, join, or assist a labor organization; to bargain collectively through representatives of their choosing; and to engage in concerted activities for mutual aid and protection. Section 7 also provides that employees have the right to refrain from any or all such activities except to the extent that such right may be affected by an agreement requiring membership in a labor organization as a condition of employment. Most of the provisions of Section 7 of the Labor-Management Relations Act are designed to protect the rights of the employee identified above.

Safety professionals should pay particular attention to the actions or inactions that constitute an unfair labor practice under Section 8 of the NLRD. Safety professionals, as agents for the company or organization, can be charged with an unfair labor practice against the company or organization if they: "(1) interfere with, restrain or coerce employees in the exercise of the rights guaranteed under Section 7; (2) dominate or interfere with the formation or administration of any labor organization or contribute financial or other support to it; (3) discriminate in regards to hire or tenure of employment or any term or condition of employment to encourage or discourage membership in any labor organization; (4) discharge or otherwise discriminate against an employee because he has filed charges or given testimony under the Act; and/or (5) refuse to bargain collectively with representatives of his employees."[1] Safety professionals should be aware of the activities that constitute an unfair labor practice, and even if they are acting in good faith, the performance of the safety function can often place safety professionals in a position where an unfair labor practice change may be filed by the employee or labor organization.

Safety professionals should be aware that many companies or organizations go to great lengths to avoid unionization. Safety professionals, as an agency of the company or organization, can easily be "caught in the middle" during a union-organizing campaign, with unfair labor practice charges being filed against the company or organization for the safety professional's actions or inactions. Some of the common unfair labor practice charges filed by employees or labor organizations include, are but not limited to, interrogation of employees; polling of employees; investigative interviews (in violation of the *Weingarten* rule); threats, promises, and reprisals;

granting of benefits; and spying. Some of the activities and statements that the NLRB could find to be in violation of Section 8(a)(1) include the following:

Threats
- A safety professional threatens an employee with violence because of the employee's union activities.
- A safety professional tells an employee, "The company will never sign a contract with that union."
- A safety professional tells an employee that the company knows who signed the authorization cards and will "get them."

Interrogation
- A safety professional asks an employee, "Who is attending the union meeting tonight?"
- A safety professional asks an employee how he or she feels about the union.
- A safety professional asks an employee how other employees feel about the union.

Promises
- A safety professional states an employee will get a raise if he or she votes against the union.
- A safety professional tells an employee he or she will get a promotion if he or she votes against the union.

Spying
- A safety professional implies surveillance when he or she tells an employee that he or she had a lot to say at the union meeting last evening.
- A safety professional visits an employee's home for the purposes of ascertaining his or her union support.

The above are only examples of possible violations of Section 8(a)(1) and should not be construed in any way as all inclusive. Safety professionals with responsibilities within the areas of workers' compensation and security, as well as safety and health, should exercise caution and acquire legal guidance before embarking into any activities or conversations that could constitute an unfair labor practice.

Safety professionals should be aware that unfair labor practices can also be committed by the labor organization under Section 8(b). Labor organizations and their agents are prohibited from restraining or coercing employees in the exercise of their Section 7 rights or employers in the selection of their representatives for the purposes of collective bargaining or grievances. Additionally, it is an unfair labor practice to cause or attempt to cause an employer to discriminate against an employee with respect to membership in an organization being denied or terminated on the grounds of failure to provide periodic dues or initiation fees uniformly required as a condition of acquiring or maintaining membership.

Safety professionals should be aware that most private sector safety professionals fall within Section 2(11) as "supervisors" and are considered management, and thus outside of the collective bargaining unit. A supervisor means "any individual having authority, in the interest of the employer, to hire, transfer, suspend, layoff, recall, promote, discharge, assign, reward, or discipline other employees, or responsible to direct them, or to adjust their grievances, or effectively to recommend such action, if in connection with the foregoing exercise of such authority is not of a merely routine or clerical nature, but requires the use of independent judgment."[2]

As a representative of management, safety professionals should exercise caution when contacted by a representative of a labor organization. Labor organizations often employ tactics in an attempt to acquire recognition by the company or organization before an election. For example, a union representative may contact the safety professional to ask for an opportunity to discuss alleged complaints regarding safety violations and other safety issues. Prudent safety professionals should contact their human resource department or legal counsel immediately if this happens. Generally, safety professionals are not required to comply with the union's request until and unless the union becomes the designated representative of the employees. If the union is not recognized by the company or organization or designated by the NLRB as the employees' representative (usually after the election), the safety professional, as the agent for the company or organization, is not required to bargain with the union on any subject, including safety and health.

Safety professionals should exercise caution whenever a union-organizing campaign is underway. Guidance should always be sought from the human resource department or legal counsel. Generally, safety professionals should not look at any lists of employees provided by the labor organization or look at any authorization cards or letters from the labor organization. This could constitute recognition of the labor organization. Additionally, safety professionals should not accept any registered mail or any written documents attempted to be handed to the safety professional from the labor representative.

A labor-organizing campaign can be very disruptive to safety and health program efforts. Safety professionals should be aware that the labor organization must acquire signed authorization cards from 30% of the employees in the identified collective bargaining unit in order for the NLRB to order an election. The NLRB conducts a secret ballot election, usually on site, and employees who sign the authorization cards are not bound by the card to vote for the union. If the majority of the employees vote for a union, the NLRB would recognize the union and order collective bargaining negotiations with the company or organization to begin. If the union does not receive a majority of the votes from the collective bargaining unit employees, the NLRB has the ability to recognize the union and order the company or organization to bargain with the union. However, before the NLRB can order bargaining, the union must show that the employer improperly denied the union's preelection bargaining requests or the union lost its majority as a result of unfair labor practices by the company or organization. If the company or organization committed serious unfair labor practices, the NLRB can order the company or organization to bargain with the union if the union obtained a majority of the authorization cards without even holding a secret ballot election.

Given the visibility and functions of the safety professional within the company or organization, the safety professional can often be a focus point for the filing of unfair labor practices by the labor organization. One of the first tactics often employed is to contact OSHA to initiate an inspection in order that any alleged violations could potentially be utilized during the labor organization's organizing campaign. The unfair labor practice charges, which usually are filed by both the labor organization and the company, play an important role in the organization process. If the labor organization can prove unfair labor practices by the company or organization, the NLRB can order the company to bargain with the union even if the union lost the election. Additionally, in the event of a strike or lockout situation, unfair labor practices play a key role in determining whether the strike was an economic strike or an unfair labor practice strike, and thus whether the striking employees or replacement employees return to the jobs.

A labor-organizing campaign can be a minefield for the safety professional. Safety and health programs, or lack thereof, may be one of the reasons why the employees initially contacted the labor organization. The organizing campaign can be very disruptive to the overall safety and health efforts within the company or organization while the company or organization and the union battle for the hearts and minds of the voting employees. Safety professionals, as a member of the management team, should exercise caution in their daily job activities and be extremely cautious in their interaction and communications with employees and the labor organization representatives. The safety professional's actions or inactions directly reflect on and impact the company or organization. Prudent safety professionals should always receive their direction and advice from their human resources or legal counsel when encountering a labor-organizing campaign.

CHAPTER QUESTION

1. The NLRA is _____.
    a. National Labor Relations Act
    b. National Labor Reunion Act
    c. National Longshoreman and Railroad Workers Act
    d. None of the above
2. A labor organization must have ____% of the authorization cards tom initiate a campaign.
    a. 20%
    b. 30%
    c. 40%
    d. None of the above
3. A violation of Section 8(a)(1) can include
    a. threats to employees.
    b. interrogation of employees.
    c. promises to employees.
    d. All of the above.
4. An *Excelsior* list includes
    a. names and addresses of union officials.
    b. names and addresses of company officials.
    c. names and addresses of employees.
    d. None of the above.

5. A "closed shop" is
   a. permitted in all states.
   b. illegal.
   c. permitted in some states.
   d. None of the above.

ANSWERS

1. a
2. b
3. d
4. c
5. b

ENDNOTES

1. NLRA, Section 8(a).
2. NLRA, Section 2(11).

4 NLRB and Labor-Management Relations

I never did anything worth doing by accident, nor did any of my inventions come by accident; they came by work.

—**Thomas Edison**

Men are not against you; they're merely for themselves.

—**Gene Fowler**

Learning objectives:

1. Acquire an understanding of the functions of the National Labor Relations Board.
2. Acquire an understanding of unfair labor practices.

"The **National Labor Relations Board** is an independent federal agency vested with the power to safeguard employees' rights to organize and to determine whether to have unions as their bargaining representative. The agency also acts to prevent and remedy unfair labor practices committed by private sector employers and unions."[1] The National Labor Relations Board (NLRB) was established by the Wagner Act in 1935 to administer and enforce the provision of the National Labor Relations Act.

The NLRB consists of three members serving 5-year terms as well as general counsel, administrative law judges, an executive secretary, and regional directors. There are currently thirty-three regional NLRB offices and several subregional offices located throughout the United States. The Labor-Management Relations Act divided the authority of the NLRB between the board and the general counsel. In 2009, President Obama designated Wilma B. Liebman as the chairman of the NLRB and nominated attorneys Craig Becker and Mark Gaston Pearce as members of the board.[2] It should be noted that the NLRB has been functioning with only two members of the board since 2008.[3]

The NLRB is primarily a judicial body hearing unfair labor practice (ULP) charges and representation matters; however, the NLRB does have limited investigative and prosecutorial authority. The office of general counsel has absolute authority over the issuance of unfair labor practice complaints. Section 6 of the NLRA provides the authority to the NLRB to issue rules and regulations related to the NLRA.

Safety professionals should note that an actual labor dispute must exist before the NLRB possesses jurisdiction, and thus can become involved in the situation. Unlike OSHA and other governmental agencies, the NLRB cannot become involved until there is a controversy involving the "terms, tenure or conditions of employment, or concerning the association or representation of persons in negotiating, fixing, maintaining, changing or seeking to arrange terms or conditions of employment, regardless of whether the disputants stand in the proximate relation of employer and employee."[4] In addition to a labor dispute, the NLRB is required to identify that the labor dispute "affects commerce."[5]

The NLRB is responsible for conducting labor organization representation elections in a "fair and impartial manner."[6] To conduct this election, the NLRB has adopted the laboratory conditions standard, which the NLRB imposes on the workplace and both the company or organization and the labor union. Laboratory conditions are usually imposed before and during the election on both the labor union and the company or organization. Safety professionals should be aware of when the NLRB's laboratory conditions are in effect, given the fact that if the laboratory conditions are violated, the NLRB can set aside the election and conduct it over again. The NLRB usually evaluates the conduct of the labor union and company representatives from the time the petition is filed by the labor organization up to and including the election.[7]

Safety professionals should be aware that historically, the NLRB has varied in its position regarding setting aside elections for misrepresentative statements about material facts.[8] In 1962, the NLRB established the *Hollywood Ceramics*[9] standard, wherein an election was to be set aside where there existed a misrepresentation of a material fact involving a substantial departure from the truth and stated at a time where there was no opportunity to reply by the other company or labor organization, and the statement had a significant impact on the free choice of employees before the vote.[10] However, in 1982, the NLRB stated that they were no longer following the *Hollywood Ceramics* standard and would set aside an election only when the company or labor organization used forged documents in misleading propaganda.[11]

Of particular interest for safety professionals is Section 8(c), which provides that management of the company and agent of the labor organization have the right to express any views or opinions; however, statements containing threats of reprisals or promises of benefits are prohibited.[12] Safety professionals should be aware that any preelection statements regarding the loss of jobs, loss of business, layoffs, or closure of the operations are generally grounds for setting aside an election.[13] Safety professionals should also be aware that elections have been set aside where labor organization agents threatened workers with physical harm if they did not vote for the labor organization.[14]

Safety professionals often visit injured employees at their homes or conduct job safety observations of employees performing their job functions. During the laboratory conditions prior to an election, safety professionals should be aware that these activities can be construed as interrogation and surveillance. If the company attempts to gain information about the labor organization's activities through interrogation or surveillance, this may be grounds to set aside the election.[15] Safety professionals should be aware that the NLRB regards visits to workers' homes during the preelection period as objectionable conduct irrespective of whether threatening or coercive comments or remarks are made during the visit.[16] The primary reason for this prohibition is the potential to subject a fear of discrimination in the minds of the workers and affect their free choice in the election.

Propaganda is often used by companies and labor organizations during the pre-election stage. Safety professionals should be aware that they may be the subject of this propaganda and the propaganda can be personal in nature. The NLRB will monitor these activities during the preelection laboratory conditions and can set aside an election if the propaganda is of a racial nature or of a substantially inflammatory nature.[17] However, safety professionals should be aware that the NLRB will not invalidate an election for such propaganda as character attacks unless coercion, fraud, or other attacks impact the laboratory conditions of the election.[18] Given the fact that many safety professionals are substantially visible in the operations, and safety is often a primary issue in the election, it is not unusual for the safety professional to receive propaganda specifically focused on the safety programs or their personal selves.

Any type of electioneering at or near the polling place where the election is being held by the NLRB can result in the election being set aside. Safety professionals should be aware that any prolonged conversation in the polling area can be sufficient to meet this requirement. Additionally, even being in the area of where the election is being held can be considered impermissible conduct and grounds to set aside an election.[19]

As can be seen, the environment in which the safety professional functions in the workplace and the activities that the safety professional can perform can be substantially modified during the laboratory conditions leading up to an NLRB election. Safety professionals should take their lead and direction as to which activities or functions should or should not be performed during this period of time from their human resources department or legal counsel. Safety professionals should be aware that activities or functions that may have constituted the norm prior to the organizing campaign can constitute an unfair labor practice or grounds to set aside an election during this period of laboratory conditions before an election. The NLRB is responsible for conducting the election by the employees in a fair and impartial manner; however, safety professionals should be aware that this does not prevent companies or labor organizations from attempting to sway the minds and opinions of the workers to vote for their side. Given the importance of safety in the workplace and the visibility of the safety professional in the operation, safety professionals should be prepared for program or personal challenges and be knowledgeable as to proscribed methods to address these challenges.

If confronted with a labor-organizing campaign, it is imperative that the safety professional seek guidance from his or her human resources department or legal counsel from the onset. Safety professionals should also research the most recent NLRB decisions on any particular issue that may arise on the NLRB website: www. NLRB.gov. Safety professionals should be aware that the "rules of the road" will change dramatically during an organizing campaign leading to an NLRB election, and safety professionals do not want to jeopardize their company's position with their workers through errors and missteps.

CHAPTER QUESTIONS (TRUE/FALSE)

1. The NLRB is a ten-member commission.
2. Visiting employees at their homes can be an ULP.
3. Threatening to close the plant can be an ULP.
4. The NLRB has no authority to invalidate an election.
5. The NLRB was established under the OSH Act.

ANSWERS

1. False
2. True
3. True
4. False
5. False

ENDNOTES

1. NLRB website: www.nlrb.gov.
2. www.nlrb.gov/shared_files/Press %20Releases/2009/B04709.pdf.
3. As the result of a deadlock between President Bush and the Senate, a third member of the board was not confirmed. The U.S. Supreme Court has agreed to consider whether the NLRB has authority to issue two-member rulings, which places the decisions made by the board during this time period at risk.
4. 29 USCA Section 152(9).
5. 29 USCA Section 159(c).
6. *Collins & Aikman Corp. v. NLRB*, 383 F.2d 722 (CA-4, 1967).
7. *NLRB v. Blades Mfg. Co.*, 344 F.2d 998 (CA-8, 1965).
8. See *Shopping Kart Food Markets, Inc.*, 228 F.2d 957(CA-6, 1981), reversing Hollywood Ceramics Co., 140 NLRB 221 (1962).
9. Id.
10. Id.
11. Midland Nat. Life Ins. Co., 263 NLRB No. 24 110 (1982).
12. NLRA Section 8(c).
13. See Vegas Village Shopping Corp., 229 NLRB 279 (1977) (loss of jobs); Whiting Mfg. Co. Inc., 258 NLRB 58 (1981) (loss of business); Andersen Cottonwood Concrete Products, Inc., 246 NLRB No. 172 (1979) (reduction in workforce and plant closure).
14. *Loose Leaf Hardware v. NLRB*, 666 F.2d 713 (CA-6, 1981).
15. ITT Cannon Electric, Div. of ITT, 172 NLRB 425 (1968) (interrogation); General Electronic Wiring Devices, Inc., 182 NLRB 876 (1970) (surveillance).
16. Peoria Plastic Co., 117 NLRB 545 (1957).
17. Sewell Mfg. Co., 138 NLRB 66 (1962); *Schneider Mills, Inc. v. NLRB*, 390 F.2d 375 (CA-4, 1968).
18. Georgia-Pacific Corp., 199 NLRB 43 (1972).
19. Belk's Department Stores of Savannah, Ga., Inc., 98 NLRB 280 (1952).

5 Collective Bargaining

When you come to a fork in the road—take it.

—Yogi Berra

No man was ever so completely skilled in the conduct of life as not to receive new information from age and experience.

—Terence

Learning objectives:

1. Acquire an understanding of the collective bargaining process.
2. Acquire an understanding of the collective bargaining issues.
3. Acquire an understanding of the results of the collective bargaining process.

Collective bargaining is defined as "an agreement negotiated between a labor union and an employer that sets forth the terms of employment for the employees who are members of that labor union. This type of agreement may include provisions regarding wages, vacation time, working hours, working conditions, and health insurance benefits."[1] Safety professionals may or may not be directly involved in the collective bargaining negotiations process, depending on your company or organization. However, safety professionals should be aware that safety is often a major issue within the negotiation process and is considered a mandatory subject for bargaining. Whether the safety professional is at the negotiation table or working behind the scenes providing safety-related information to the spokesperson, he or she is often a key member of the negotiations team.

The initial issue that safety professionals should be aware of is the status of the labor organization. In companies or organizations where the labor organization was already present, the situation may involve a collective bargaining agreement that is set to expire and is being renegotiated. In other situations, labor organizations may have been recognized by the company or organization upon request or before the certification of the results of an NLRB election. The most common scenario in which the labor organization is recognized and negotiations are required is after a secret ballot election where certification of the results of the election identify the labor organization as the representative of the employees within the bargaining unit.

Section 8(b) of the NLRA provides the duty to both the company or organization and the labor organization to bargain collectively, and includes the mutual obligation to meet at reasonable times and to negotiate in good faith over the subjects of wages, hours, and other terms and conditions of employment, which includes safety. This section also identifies that the company and labor organization are required to execute a written contract, including any and all agreements reached by the parties.

However, Section 8(d) does not require or compel the company or labor organization to agree to the other side's proposals or to make concessions in order to reach an agreement.

Safety professionals should be aware that Section 8(a)(3) identifies that it is an unfair labor practice for the company or organization to refuse to bargain with the recognized labor organization. And under Section 9(a), the labor organization who won the majority of the votes in the election is the exclusive representative of the employees within the identified collective bargaining unit, and the company or organization is required to negotiate in good faith over wages, hours, and other conditions of employment. It should be noted that Section 9(a) does not prevent individual employees from taking grievances directly to the safety professional, as a representative of the company, as long as these communications are not inconsistent with the terms of the collective bargaining agreement.

Under Section 8(a)(5), a company or organization can commit a violation if the company or organization refuses to negotiate with the recognized union over the mandatory bargaining subjects of wages, hours, and conditions of employment.[2] This violation can be applicable even if the company or organization bargains in good faith and desires to reach an agreement. A company or organization is required to bargain only over the mandatory bargaining issues of wages, hours, and conditions of employment, and is not required to bargain over any subject or issue that is not considered mandatory.

Although the lines between the mandatory subject and a remote or nonmandatory subject can often get blurred, safety professionals should be aware that safety rules have been found to be a mandatory subject requiring negotiations.[3] Other safety-related issues that have also been found to be mandatory subjects include bonuses or gifts,[4] job transfers,[5] work schedules,[6] and antidiscrimination clauses.[7] Safety professionals should be aware that although negotiations are not limited to mandatory subjects, insistence by the company or organization or labor organization to include nonmandatory subjects in the collective bargaining agreement as a condition of agreement or signing can constitute an unfair labor practice.[8]

Under Section 9(a), safety professionals should be aware that labor organizations possess the same duty to bargain in good faith. Labor organizations also have the same duty to meet and confer with the company or organization and the same duty to bargain in good faith with the desire to reach an agreement. A labor union can commit an unfair labor practice by refusing to bargain to impasse, refusing to sign a negotiated agreement, or demanding to negotiate for a unit that has not been recognized. Safety professionals should be aware that it is also an unfair labor practice for a labor organization to bargain to impasse over nonmandatory subjects of bargaining.[9]

As can be seen, there are a multitude of different rules and regulations governing the collective bargaining process, and the requirements may change depending on the circumstances. The negotiations can take a variety of forms, from first-time negotiations, to the continuation of an existing collective bargaining agreement to a successor employer who purchased the company or organization, to a multiemployer bargaining unit. Most safety professionals will work behind the scenes to assist the spokesperson or legal counsel who is representing the company or organization at the negotiation table with specific safety or related issues. However, it is important that the safety professional be properly prepared to provide guidance and assistance before, during, and after this negotiation process.

In preparing for pending negotiations, safety professionals should be cognizant that any adversarial climate that developed between the company or organization and the labor organization before and during the election process usually continues at the negotiating table. This is often the case with initial contract negotiations where the company or organization and the labor organization were combative during the organizing campaign, with the company or organization losing the election. The tactics often employed by the company as well as the labor organizations during the organizing campaign create a level of animosity on both sides that carries over to the negotiating table.

Safety professionals should be aware that preparing for collective bargaining negotiations takes substantial time, research, training, and resources. Safety professionals should be aware that until a collective bargaining agreement is reached and ratified, supervisors and management team members retain all management authority they possessed prior to the union certification. Although the working environment should be status quo, in many circumstances the organizing campaign and election results create change in the workplace. Safety professionals should be aware that this changing work environment may require the supervisors and members of the management team to undergo significant behavioral, ego, practical, and attitudinal changes that require substantial retraining.

In preparing for the collective bargaining negotiations, safety professionals may be asked to prepare and share information with the labor organization. Although all requests should be approved by the lead negotiator or legal counsel prior to the safety professional providing any information to the labor organization, the Taft-Hartley amendments to the NLRA and orders from the NLRB can permit the labor organization to request specific information from the company or organization prior to the negotiations. Safety professionals should be aware that information often requested by labor organizations, in addition to wage data, hours worked, EEO data, and related information, includes safety inspections, safety audits, safety compliance programs, injury and illness data, training records, workers compensation data, and insurance information.

Safety professionals should be prepared to conduct a substantial amount of research, case studies, and case and data analysis for each safety or related issue in preparing for the negotiations. The information acquired is often reviewed and assembled into proposals and counterproposals for each identifiable issue and submitted to the labor organization at the bargaining table. Conversely, the labor organization will conduct its own research and develop proposals and counterproposals for each identifiable issue. Each proposal or counterproposal should be carefully analyzed with an appropriate justification or rationale being prepared to justify the proposal or counterproposal. Proposals and counterproposals should be carefully worded and provided in clear language, as well as being consistent with the position and strategy of the company or organization.

One of the most difficult challenges for safety professionals is the preparation of various analyses of data to determine the estimated cost of each of the various proposals or counterproposals. For safety professionals, the cost of various safety efforts can possess wide variables, often including, but not limited to, equipment costs, replacement costs, training costs, overtime costs, insurance costs, and related costs. For larger companies or organizations, simply the training costs of x number of employees multiplied by the employee's rate of pay multiplied by the number of training hours can reach a substantially large amount of money.

The collective bargaining process usually involves four basic steps: the initiation of the actual bargaining, the establishment of the rapport between the negotiators, the negotiation and consolidation of the proposals, and the completion of bargaining and consolidation of the agreements into the collective bargaining contract. The safety professional, no matter whether he or she is seated at the negotiating table or working behind the scenes, is an instrumental member of most negotiating teams.

Safety professionals should be well prepared at all times and be ready at a moment's notice to produce information to support any safety-related proposals or counterproposals. Safety professionals should also be aware that the negotiating process can be a grueling and time-consuming process, often stretching for weeks or even months. The location of the negotiation sessions is often away from the operations; thus safety professionals actively involved as part of the negotiating team should remember to make appropriate arrangements within the company or organization to fulfill the required duties and responsibilities in the safety professional's absence.

Safety professionals should be aware that the results of the collective bargaining negotiations can move in various directions. Although both the labor organization and company or organization possesses a duty to bargain in good faith to impasse, sometimes an impasse is reached resulting in a lockout of employees in the bargaining unit by the company or organization, usually followed by the replacement of the employees, or the labor organization can call for a strike by employees in the bargaining unit. During this time period, safety professionals should exercise caution because the intensity of the situation will increase dramatically, and often the number of unfair labor practice charges filed by the company or organization and the labor organization increases dramatically. After the conclusion of the strike, in general, if the strike is found to be purely economic, the replacement workers usually maintain their jobs. If the strike is found to be the result of an unfair labor practice by the company or organization, the striking employees usually return to the job and the replacement workers lose their jobs.

In most negotiations, an agreement between the company or organization and the labor organization is reached before impasse. The results of the various negotiated proposals and the agreed upon wages, hours, and conditions of employment that were negotiated are codified in a document, usually called the collective bargaining agreement or union contract. Section 8(b) of the NLRA provides a duty to the company or organization, as well as the labor organization, to include "the execution of a written contract incorporating any agreement reached if requested by either party."[10] If the labor organization or company or organization refuses to sign the written contract containing the terms, conditions, and provisions that were agreed to between the company or organization and the labor organization, this refusal would constitute a violation of Section 8(b)(3) of the NLRA.

Safety professionals should be aware that once signed and ratified, the collective bargaining agreement becomes the "law of the workplace." Most collective bargaining agreements require some form of grievance process wherein any alleged violation of the collective bargaining agreement can result in a grievance being filed by an employee or the labor organization, resulting in hearings up to and including arbitration. Safety professionals should carefully analyze the collective bargaining agreement and appropriately adjust or modify existing safety programs, policies, and procedures to adhere to the terms and conditions negotiated in the collective bargaining agreement. Safety professionals should be aware that any disagreements with regards to virtually any aspect of the safety and health programs, policies, or procedures usually result in formal grievances being filed by the labor organization or individual employees.

Although this chapter is a "broad brush" overview of the various aspects of the collective bargaining process and the areas that impact many safety professionals, it in no way encompasses all of the various laws and regulations that are involved in the NLRA and related labor laws. Safety professionals are not usually expected to be labor law experts, and are seldom the primary negotiators in the collective bargaining process; however, they are often key members of the negotiation team. Safety and health, as a condition of employment, are a mandatory subject of bargaining in any collective bargaining negotiations and are a vital issue in any company or organization. Safety and health are not only an important issue for safety professionals, but also the highest priority for employees and their labor representatives.

CHAPTER QUESTIONS (TRUE/FALSE)

1. Collective bargaining is a method of achieving a lower unit price.
2. A mandatory subject includes wages.
3. Safety is considered a condition of employment.
4. There must be an impasse before a strike or lockout.
5. Safety bonuses must be part of the collective bargaining negotiations.

ANSWERS

1. False
2. True
3. True
4. True
5. True

ENDNOTES

1. Definition found at lectlaw.com.
2. *NLRB v. Ins. Agents' Int. Union*, 361 U.S. 477 (1960).
3. *NLRB v. Miller Brewing Co.*, 408 F.2d 12 (CA-9, 1969).
4. *NLRB v. Electric Steam Radiator Corp.*, 321 F.2d 733 (CA-6, 1063).
5. *Rapid Roller Co. v. NLRB*, 126 F.2d 452 (CA-7, 1942).
6. *Oughton v. NLRB*, 18 F.2d 486 (CA-3, 1941).
7. *United Packinghouse, Food & Allied Workers Int. Union v. NLRB*, 135 App. DC 111, 416 F.2d 1126 (1969).

8. *NLRB v. Wooster Div. of Borg-Warner Corp.*, 356 U.S. 342 (1958).
9. NLRA Section 8(b)(3).
10. NLRA Section 8(b).

6 Contract Administration

Doing little things well is a step toward doing big things better.

—Harry F. Banks

If you find a path with no obstacles, it probably doesn't lead anywhere.

—Frank A. Clark

Learning objectives:

1. Acquire an understanding of the requirements and duties involved in contract administrations.
2. Acquire an understanding of the grievance process.

In the event of the signing and ratification of a collective bargaining agreement, contract administration is often the next step in the collective bargaining process that impacts the functions and activities of the safety professional. Safety professionals should be aware that the animosities and possible anger from the propaganda and tactics used during the organizing campaign can spill over after the collective bargaining agreement is finalized. Safety professionals should also be cognizant that the significant changes in the workplace resulting from the employees selecting a labor organization to represent them, as well as the changes resulting from the negotiated collective bargaining agreement, may create a level of challenge from the perceived deterioration of authority within the management ranks. However, once the collective bargaining agreement is reached, it is the responsibility of the labor organization and company to develop a process to administer the framework and various provisions of the collective bargaining agreement within the operations. Contract administration, although seldom the direct responsibility of the safety professional, usually impacts the performance and structure of the safety and health function.

Contract administration is usually defined as involving the establishment of rules and regulations to implement the collective bargaining agreement and settlement of grievances. However, safety professionals should be aware that most contract administration is a complex matter, and responsibilities are extensively broader than simply implementing the contract and grievances. After negotiations and agreement on the terms and conditions established in the collective bargaining agreement, the terms and conditions agreed upon are usually introduced to the workers. Many workers vote for the concept of representation by the labor organization; however, they may be unaware of the actual implementation of the terms and conditions of the collective bargaining agreement. Additionally, for safety professionals, it is important that the management team acquire the necessary skills and behaviors necessary to function within this modified workplace where there are now specific requirements and a third party overseeing their activities.

Safety professionals may also need to update their skill set and adjust their behaviors in order to comply with the requirements and structure of the collective bargaining agreement. The safety function operating under a collective bargaining agreement is substantially different from that within a nonunion environment. Safety professionals should become knowledgeable in all aspects of the collective bargaining agreement and the procedures established to adhere to the terms and conditions of the agreement in order to avoid the trial-and-error method that can result in grievances, as well as a considerable expense to the company.

Safety professionals should be aware that management within the company or organization usually plays a pivotal role within contract administration. The labor organization usually adopts a watchdog role, focusing on the company's violations of contract provisions and ensuring that management adheres to the agreed upon terms and conditions. Safety professionals, as a member of the management team, often become the contract specialists in the administration of the provisions addressing safety and health in the workplace. Safety professionals should become fully knowledgeable in the safety and health requirements within the collective bargaining agreement, as well as the perimeters of management's prerogative within the safety function.

It is not unusual for provisions of the collective bargaining agreement to require interpretation through the grievance process, or even negotiations over issues that may be unclear in the collective bargaining agreement. Safety professionals should be aware that the purpose of the grievance process is to provide an internal method to resolve conflicts, to facilitate communications between the labor organization and company, to provide a method for employees to complain, and to enforce the provisions of the collective bargaining agreement. Safety and health issues are often one of the areas of complaint by employees, and thus safety professionals should be prepared to address employee grievances within the contract administration structure.

Grievance procedures are usually established within the collective bargaining agreement and can vary in nature from contract to contract. Commonly, grievance procedures provide several internal steps of review and decision, followed by an external arbitration process. Many grievance processes start with an oral or written employee complaint to the union business agent or representative of the labor organization identifying an alleged violation of the contract, arbitrary application of contract provisions by management, or other issues involving the workplace. In rare circumstances, a large number of employees can file a class type grievance for an entire group of employees. Safety professionals should be aware that virtually all grievance procedures are one way, in that only employees covered under the collective bargaining agreement may file a grievance—management cannot file a grievance.

In this initial step of many grievance procedures, after the employee files the oral or written grievance, the business agent and representative of the company often attempt to resolve the grievance issue internally in an efficient and cost-effective manner. The next step is often a more formal hearing procedure where the issues are presented to a designated member of management and a designated representative of the labor organization. Again, the goal of this step within the grievance process is to efficiently and effectively resolve the issue on an internal basis. Safety professionals should be aware that this step may involve more than two levels of grievance review and assessment, with the early grievance procedures being conducted internally. For many grievance processes, the highest level of appeal is to an outside arbitrator or panel of arbitrators selected, agreed upon, and paid by the labor organization or company under the terms and conditions of the collective bargaining agreement.

Safety professionals should be aware that time limitations are often a strictly enforced component of the grievance process. Many grievance procedures establish specified time periods from the time of the alleged infraction or issue involved in the grievance to the filing of the oral or written grievance. If the employee(s) does not file the grievance in the manner proscribed and within the specified time periods, the grievance can often be found to be barred or become a nullified issue. This is especially important as the grievance moves higher in the process. The time period allotted for documents, witnesses, or other material evidence to support or rebut the issue at hand is usually limited in order that the grievance process can move in an efficient manner. Safety professionals involved in the grievance process should become knowledgeable in the time limitations established in the collective bargaining agreement or grievance procedures.

Given the fact that safety and health impact virtually each and every worker in virtually every operation, safety professionals should be aware that it is not unusual for safety and health to be the topic of numerous grievances, ranging from individual safety issues, such as personal protective equipment (PPE) selection, to broader program issues, such as the implementation of a new compliance program. Safety professionals should not take the grievances filed by employees lightly and should properly prepare for each and every grievance meeting or hearing appropriately. It is important for safety professionals to always act in a professional manner in grievance meetings and hearings, even if the arguments become heated or personal in nature.

Safety professionals should remember the primary purpose of the grievance process is to find a peaceful resolution to the issue while maintaining the conflict internally within the grievance process or before an impartial decision maker. Safety professionals should be prepared to defend their position in a vigorous but professional manner and remember to respect the employee and always treat him or her with dignity. And safety professionals should remember that the goal is to amicably settle the dispute; thus safety professionals should be prepared to compromise, where necessary, to achieve a win-win situation where possible. The grievance process is part of the overall contract administration; however, the safety professional can play a pivotal role in achieving a safe and healthy workplace, as well as a workplace where employees are valued and respected for their work efforts.

CHAPTER QUESTIONS (TRUE/FALSE)

1. Contract administration includes negotiation of safety rules.
2. Management usually controls the contract administration process.
3. Grievance procedures are usually negotiated in the collective bargaining agreement.
4. A grievance procedure often includes arbitration.
5. A grievance procedure often includes mediation.

ANSWERS

1. True
2. True
3. True
4. True
5. False

7 Title VII and Antidiscrimination Laws

No task is so humble that it does not offer an outlet for individuality.

—Wililam Feather

What counts in any system is the intelligence, self-control, conscience and energy of the individual.

—Cyrus Eaton

Learning objectives:

1. Acquire an understanding of the requirements of Title VII of the Civil Rights Act.
2. Acquire an understanding of the functions of the Equal Employment Opportunity Commission.
3. Acquire an understanding of the unlawful employment practices under Title VII.

Although the prevention of discrimination in the workplace is usually not within the direct responsibilities of most safety professionals, discrimination in the workplace can be an area in which unknowing safety professionals can "step on their tail" very quickly in their daily activities. The scope of antidiscrimination laws is substantial, encompassing not only federal laws but also state and even local laws and regulations. The purpose of virtually all antidiscrimination laws is to "remove artificial, arbitrary and unnecessary barriers when such impediments operate invidiously to discriminate against individuals."[1] The areas covered by the myriad federal, state, and local laws include race, sex, color, creed or religion, national origin, pregnancy, age, disability, equal pay/compensation, retaliation, genetic information, and sexual harassment.

Safety professionals should be aware that on a federal level, the Equal Employment Opportunity Commission (EEOC) is charged with administering and enforcing the various antidiscrimination laws. Many states possess correlating state agencies to administer and enforce the state laws. The EEOC is composed of five members appointed by the president and confirmed by the U.S. senate for a term of 5 years. The EEOC possesses an office of general counsel to prosecute violations and has regional or district offices located throughout the United States. The EEOC also compiles and publishes data that, along with information about its services and laws, is available on its website: www.EEOC.gov. Safety professionals should be aware that there are a number of other federal governmental agencies, including the U.S. Department of

Labor (which includes OSHA) and the Immigration and Naturalization Service, that also enforce antidiscrimination laws.[2]

The primary federal law that most safety professionals are aware of is Title VII of the Civil Rights Act of 1964,[3] commonly referred to as Title VII. This law was the most sweeping civil rights legislation ever enacted and contains eleven different titles addressing discrimination in public accommodations, voting, education, and most important to safety professionals, in the employment setting. Safety professionals should be aware that the Civil Rights Act has been modified and amended on several occasions since 1964, with the most recent being the Civil Rights Act of 1991.[4] In the Civil Rights Act of 1991,[5] Congress reacted to a series of court decisions by the U.S. Supreme Court changing the landscape of discrimination law and precedent. With the Civil Rights Act of 1991, Congress amended several of the statutes enforced by the EEOC and added jury trials, compensatory and punitive damages in Title VII (and ADA) lawsuits involving intentional discrimination, and statutory caps on damages awarded for future losses, pain and suffering, and punitive damages, depending on the size of the employer. Additionally, Congress codified the disparate impact of the theory of discrimination and expanded the coverage of Title VII to cover employees of American-controlled companies or organizations with operations outside of the United States.

Safety professionals should be aware that Section 708 of Title VII permits each state to have parallel state legislation and regulation in the employment setting as long as the law does not conflict with Title VII. Today, every state possesses a law prohibiting discrimination in the employment setting. Although virtually all companies or organizations employing safety professionals fall within the jurisdiction of Title VII, safety professionals should be aware that employers with fifteen or fewer employees, and employees employed for each workday in twenty or more calendar weeks in a calendar year and engaged in interstate commerce, can be considered outside of Title VII's jurisdiction.[6]

For most safety professionals working in the private sector, Title VII identified unlawful employment practices to:

- fail or refuse to hire or to discharge any individual, or otherwise to discriminate against any individual with respect to his or her compensation, terms, conditions, or privileges of employment, because of such individual's race, color, religion, sex or national origin; or
- limit, segregate, or classify his or her employment or applicants for employment in any way which would deprive or tend to deprive any individual of employment opportunities or otherwise adversely affect his or her status as an employee, because of such individual's race, color, religion, sex or national origin.[7]

In addition to employers, safety professionals should be aware that labor unions, employment agencies, and joint labor-management training committees are also required to comply with Title VII. For employers, labor organizations, and others covered under Title VII, retaliation is prohibited in any manner. Safety professionals should be aware that it is considered an unlawful employment practice "to discriminate against the respective employees, applicants, members, or other related individuals because of their opposition to an unlawful employment practice or because of their filing a charge, testifying, assisting, or participating in any manner in an investigation, proceeding, or hearing under the Act."[8]

Safety professionals should become familiar with the scope and application of the protections extended to individuals under Title VII. Within the categories of race and color, Title VII includes protections to all races, including Caucasians.[9] In the category of national origin, safety professionals should be aware that language requirements are often found to be improper unless job relatedness can be shown,[10] and prohibiting employees of foreign descent from using their native language at work has also been found to be an unlawful employment practice unless a business necessity can be proven by the employer.[11] Title VII prohibits discrimination against male or female individuals based on their sex.[12] Safety professionals should be aware that the prohibition against discrimination based on sex extends beyond the hiring process, to include such other areas as life insurance programs, profit sharing plans, bonus programs, and all other plans or programs.[13]

Of particular importance for safety professionals to recognize is the area of sexual harassment. The EEOC has established special obligations for companies or organizations to ensure that their supervisory personnel, as well as other employees, do not engage in the unlawful employment practice of sexually based harassment. Safety professionals should be aware that sexual harassment can include "any unwelcome sexual advances, requests for sexual favor, or other verbal or physical conduct of a sexual nature."[14] Additionally, safety professionals should be aware that sexual advances or requests for sexual favors in the workplace substantially and detrimentally affect an employee's work performance, or can create an "intimidating, hostile or offensive work environment."[15] Safety professionals should be aware that petty slights, annoyances, and isolated incidents (unless extremely serious) will not rise to the level of illegality. To be unlawful, the conduct must create a work environment that would be intimidating, hostile, or offensive to reasonable people. Offensive conduct may include, but is not limited to, offensive jokes, slurs, epithets or name calling, physical assaults or threats, intimidation, ridicule or mockery, insults or putdowns, offensive objects or pictures, and interference with work performance. Harassment can occur in a variety of circumstances, including, but not limited to, the following:

- The harasser can be the victim's supervisor, a supervisor in another area, an agent of the employer, a coworker, or a nonemployee.
- The victim does not have to be the person harassed, but can be anyone affected by the offensive conduct.
- Unlawful harassment may occur without economic injury to, or discharge of, the victim.[16]

Safety professionals should be aware that Title VII defines *religion* as including all aspects of religious observances, practices, or beliefs.[17] The EEOC expanded this definition in its regulations to include moral and ethical beliefs not confined to theistic concepts or to traditional precepts that are sincerely held by individuals with the strength of traditional religious views and beliefs.[18] Safety professionals should be aware that Title VII goes beyond simple neutrality in the workplace, providing for companies or organizations to provide reasonable accommodations of an employee's religious observance or practice.[19] However, Title VII also requires the employee seeking to observe his or her religious beliefs to provide proper notification to the company or organization as to his or her religious needs, and an obligation to resolve any conflicts between job requirements and religious observances or practices.[20] Safety professionals should be aware that requiring an applicant or employee to wear clothing or apparel other than the clothing required by the employee's religion may constitute an unfair employment practice.[21]

In the category of pregnancy discrimination, safety professionals should be aware that the Pregnancy Discrimination Act was added as an amendment to Title VII. Safety professionals should be aware that unlawful employment practices in this category can include exclusion from medical or insurance programs, denial of a leave of absence, or discrimination based on the time or duration of a leave of absence. Of particular note is that protections under the Pregnancy Discrimination Act extend not only to female employees, but also to the spouses of male employees.[22]

Safety professionals should be aware that claims of discrimination often fall within the primary theories of disparate impact, disparate treatment, or hostile work environment. The disparate impact theory alleges a situation or activity where the discriminator impact of a policy, program, or activity affects the individual in a discriminatory manner. The disparate treatment theory involves claims alleging that similarly situated individuals are treated differently because of their race, color, or national origin. In general, a hostile work environment exists when an employee experiences workplace harassment and fears in the workplace because of the offensive, intimidating, or oppressive atmosphere generated by the supervisor, other employees, or others in the workplace. Safety professionals should be aware that the company or organization possesses an affirmative duty to address any allegations of discrimination in the workplace. If a safety professional observes or is informed of any situation involving possible discrimination in the workplace, the safety professional should report this information to the human resource department, legal counsel, or other appropriate reporting official in the company or organization.

Safety professionals should be aware that most companies or organizations possess policies addressing discriminatory practices in the workplace and often establish specific methodologies through which to report alleged instances of discrimination. The EEOC requires companies and organizations to post and keep posted a notice prepared by the EEOC identifying employees' rights under Title VII, prohibited practices, and contact information. Additionally, many companies and organizations are required to maintain specific personnel records and data, such as the EEO-1 form, which is annually reported to the EEOC.

In many situations, when an employee notifies the company or organization regarding the alleged instance of discrimination, an investigation is conducted and corrective action, if necessary, is taken internally to resolve the alleged discriminatory situation. However, safety professionals should be aware that employees may file claims with the EEOC or parallel state agencies as well as pursuing legal action under many state laws. Focusing on the EEOC, an employee possesses 180 days from the occurrence of the alleged discriminatory act to file a charge.[23] An EEOC charge must be in writing, identifying the company or organization against whom the allegations are directed, the alleged unlawful conduct must be specified, and the charge signed and verified.[24] If the claim is filed in a timely manner, the EEOC notifies the company or organization in writing of the claim and a fact-finding conference is usually held within a short time period. Safety professionals should be aware that the EEOC often requests documents and data related to the charge prior to the fact-finding conference, or the EEOC investigators can make an on-site visit to the company or organization's worksite.

Upon completion of the investigation by the field agents, safety professionals should be aware that the EEOC has to make an assessment as to the credibility of the charge(s). At this point, the EEOC usually either elects to pursue the charges if it believes the charges are true or issues a no-cause finding. Safety professionals should be aware that a no-cause determination requires the EEOC to send the employee a right-to-sue notice and advises the employee that the EEOC will dismiss the charges. The right-to-sue notice advises the employee that if he or she so desires, he or she can pursue an action in district court under Title VII, and the action must be initiated in 90 days.[25] Given the current backlog of charges, safety professionals should also be aware that if the EEOC is not able to investigate after 180 days from the date of the charge, it can provide a right-to-sue notice to the employee upon written request.[26]

Safety professionals should be aware that if the EEOC finds cause, the company or organization is often invited to participate in the conciliation process facilitated by a trained conciliator to attempt to resolve the issue(s). Safety professionals electing to pursue this option should be aware that the EEOC requires certain conditions be included and specific clauses be included in any agreement.[27] If a settlement can be achieved, the employee, company, or organization and the EEOC enter into a written conciliation contract addressing the terms and conditions of the agreement and signed by the parties.[28] If the written conciliation agreement is approved by the conciliator's bosses, a fully executed copy of the agreement is provided to the employee and the company or organization, and a copy files with the EEOC's office in Washington, D.C. If a settlement cannot be achieved or the employee or company refuses to participate in the conciliation, the EEOC will send a written notice to the parties as to its determination.

A Title VII action is considered equitable in nature, and the primary objective is to stop or enjoin the unlawful discrimination practices. Potential remedies can include injunctive relief, back pay, attorney fees, punitive damages, and other relief. Class actions are permitted under Title VII if the rules of civil procedure are met. Although safety professionals will seldom be involved in a Title VII court action, there are a number of possible defenses, including, but not limited to, business necessity defense, statistical data, and bona fide occupational qualifications.

It is vital that safety professionals recognize the areas and issues in which potential discriminatory practices could take place in the workplace and take a proactive stance to avoid any type of discrimination in the workplace. Safety professionals should strive to always be fair and equitable in all situations or decisions and always comply with the requirements of Title VII and other federal and state laws addressing discrimination in the workplace. If the safety professional possesses any shred of doubt in his or her mind, consultation with the human resources department or legal counsel is strongly advised before addressing the situation.

CHAPTER QUESTIONS (TRUE/FALSE)

1. The Department of Labor is the governing federal agency for Title VII.
2. Title VII prohibits discrimination in the area of religion.
3. Title VII covers only employees of a company.
4. The Pregnancy Discrimination Act was invalidated in 2009.
5. A violation of Title VII must be reported within 60 days of the occurrence.

ANSWERS

1. False
2. True
3. False
4. False
5. False

ENDNOTES

1. See *Griggs v. Duke Power Co.*, 401 U.S. 424 (1971).
2. These laws include the Civil Service Reform Act, Immigration Reform and Control Act, Executive Order 11246, certain sections of the Americans with Disabilities Act and Civil Rights Act of 1964, Family and Medical Leave Act, Occupational Safety and Health Act, and sections of the Rehabilitation Act, Social Security Act, Fair Labor Standards Act, and workers' compensation law.
3. 42 USCA Section 2000E through 2000E-17.
4. Public Law 102-166 (1991).
5. Ibid.
6. 42 USCA Section 2000e(b).
7. 42 USCA Section 2000-2(a).
8. 42 USCA Section 2000e-2(a).
9. This is often referred to as reverse discrimination actions.
10. EEOC decision 73-0377 (1972).
11. *Saucedo v. Brothers Well Service, Inc.*, 464 F.Supp. 919 (DC Tex., 1979).
12. 42 USCA Section 2000e-2.
13. 29 CFR Section 1604.9.
14. EEOC, Sex Discrimination Guidelines, 29 CFR Section 1604.11.
15. 29 CFR Sections 1604.2 and 1604.11.
16. EEOC website: www.eeoc.gov.
17. 42 USCA Section 2000e(j).
18. 29 CFR Section 1605.1.
19. 42 USCA Section 2000e(j); 29 CFR Section 1605.2.
20. *Chrysler v. Mann*, 561 F.2d 1282 (CA-8, 1977).
21. EEOC decision 71-2620 (1971).
22. *Newport News Shipping & Dry Dock Co. v. EEOC*, 32 FEP 1 (S.Ct. 1983).
23. 42 USCA Section 2000e(5)e.
24. 29 CFR Section 1601.8.

25. 29 CFR Section 1601.29(e).
26. 29 CFR Section 1601.24(a); EEOC Compliance Manual, Section 64.
27. EEOC Compliance Manual.
28. 29 CFR Section 1601.24(a).

8 Age Discrimination

The more men, generally speaking, will do for a dollar when they make it, the more that dollar will do for them when they spend it.

—**William J. H. Boetcker**

Frugality is founded on the principle that all riches have limits.

—**Edmund Burke**

Learning objectives:

1. Acquire an understanding of the issues involved in the aging of the American workforce.
2. Acquire an understanding of the Age Discrimination in Employment Act.
3. Acquire an understanding of the prohibited practices under the ADEA.
4. Acquire an understanding of the Older Workers Benefits Protection Act.

With the aging American workforce, employees are "aging gracefully" and remaining in the workplace longer, by either choice or economic necessity. Safety professionals now and in the near future will encounter new and different challenges in safeguarding the aging worker within the private sector workplace. The physical skills set as well as cognitive skill level may diminish with time; however, the experience and expertise the older worker brings to the workplace is often invaluable. With this changing workplace, safety professionals should be aware of the various federal and state laws that offer protection to the older employee against discrimination in the workplace based upon his or her age.

As noted in the article titled "Discrimination and the Aging American Workforce: Legal Analysis and Management Strategies," "The increasing age of the workforce, the presence of age bias in society generally, together with the fact that the consequences of unemployment fall more harshly on older people, make the topic of age discrimination a very significant one—legally, ethically, and practically."[1] Moreover, as "older" employees get even older, their pension and health care costs concomitantly increase for their employers, thereby making the older employees more attractive targets for workforce downsizing. Furthermore, older employees are disadvantaged in their efforts not only to retain employment, but also to regain employment when they are discharged from their jobs. Weak economies today also adversely affect older workers more harshly, particularly since when business is not good, employers may feel compelled to reduce the number of their most expensive employees, who are typically their oldest workers. Moreover, in a tight economy, older workers are the ones most likely to have a more difficult time to secure a job, let alone a comparable job, after they are downsized.[2]

The primary law on the federal level offering protections to older workers is the Age Discrimination in Employment Act of 1967 (ADEA).[3] "The purpose of the ADEA is to promote employment of older workers between the ages of 40 and 70 years of age based on their ability, rather than age, and to prohibit arbitrary age discrimination in employment.[4] Similar to Title VII, the ADEA prohibits discrimination by companies or organizations, labor unions, and employment agencies against employees, members, referrals, and applicants.[5]

Safety professionals should be aware that the ADEA identified the following practices by companies or organizations as being unlawful:

- To fail or refuse to hire or to discharge any individual or otherwise discriminate against any individual with respect to his or her compensation, terms, conditions, or privileges of employment because of such individual's age
- To limit, segregate, or classify its membership, or to classify or fail or refuse to refer for employment any individual, in a way that would tend to deprive that individual or employment opportunities, or which would limit such employment opportunities or otherwise adversely affect his or her status as an employee or as an applicant for employment, because of such individual's age
- To cause or attempt to cause an employer to discriminate against an individual in violation of the act.[6]

The administration and enforcement of the ADEA is currently housed with the EEOC on the federal level. This has not always been the case. The ADEA was originally administered and enforced by the Wage and Hour Division of the U.S. Department of Labor, and the Fair Labor Standards Act structure and mechanisms were utilized. With the Reorganization Plan No. 1 in 1978, the Equal Employment Opportunity Commission was provided authority over the ADEA and established its own guidelines. However, the ADEA must still be enforced through the procedures established under the Fair Labor Standards Act rather than those of Title VII, with other similar federal antidiscrimination laws.[7]

Safety professionals should be aware that the ADEA was amended in 1990 by the Older Workers Benefit Protection Act (OWBPA). Although the primary purpose of the OWBPA was to provide additional protections in the area of employee benefits, the OWBPA also addresses the issue of waivers of rights by protected older individuals. The OWBPA amended the ADEA to specifically prohibit companies or organizations from denying benefits to older employees. However, the OWBPA does, in specific situations, permit companies or organizations to reduce benefits based on the employee's age as long as the cost of providing the reduced benefits is the same as the cost of furnishing benefits to younger employees. In the area of waiver, safety professionals should be aware that the OWBPA contains very specific provisions through which older workers may waive their rights to sue or waive their rights to benefits. Specifically, the OWBPA requires that any waiver must:

1. Be in writing and understandable
2. Specifically refer to ADEA rights or claims

3. Not waive future rights or claims
4. Be in exchange for valuable consideration
5. Advise the person in writing to consult with an attorney before signing the waiver
6. Provide the person with at least 21 days to consider the waiver or agreement and at least 7 days to revoke the agreement or waiver after signing it.[8]

Safety professionals should be aware that the ADEA originally offered protection for individuals between the ages of 40 and 65 years of age, but the upper age limit was changed to 70 years of age. In 1986, Congress removed the upper age limit of 70 almost entirely from the ADEA, thus providing protection for individuals age 40 and older. The lower age limit of 40 years of age has not been addressed by Congress and remains the same in the ADEA; however, prudent safety professionals should investigate the protected ages in their state statutes, which can be significantly different than that of the ADEA.

Safety professionals should be aware that an individual alleging a violation of the ADEA can file a charge with the EEOC or bring a civil action in an appropriate court of law. Similar to the EEOC requirements for Title VII, the EEOC requires the charge to be filed within 180 days of the date of the alleged unlawful act unless the unlawful act occurred in a state that possesses a similar state agency that addresses state age discrimination laws, where the time limit is usually 300 days. The ADEA charge, again like Title VII, must be in writing, identify the company or organization, and describe the alleged conduct in violation of the ADEA.[9] Again, similar to the requirements of Title VII, companies or organizations are required to maintain records regarding benefits plans, merit systems, and preemployment applicants.[10]

Under the ADEA, an individual is entitled to file a civil action in federal district court after waiting 60 days after filing with the EEOC if the following general conditions are met:

- Sixty days has elapsed since the filing of the charge with the EEOC or state agency.
- The age discrimination charge was filed within the 180-day time limitation (or 300 days with a state agency).
- The civil action is filed within the 2- or 3-year statute of limitations.[11]

In a civil action, the *prima facie* case that the individual alleging age discrimination must establish generally includes the following:

- The individual is in an age-protected class under the ADEA.
- The individual applied for and was qualified for the position or promotion for which the company or organization was seeking applicants.
- The individual suffered an adverse employment action (e.g., discharge, demotion, etc.).
- After the individual's rejection, discharge, or demotion, the position remained open and the company or organization continued to seek applicants from people with the same qualifications as the individual.

With ADEA charges or lawsuits, safety professionals should be aware that, in addition to direct or "smoking gun" evidence of age discrimination, individuals alleging age discrimination may utilize the disparate treatment theory, disparate impact theory, pretext, and circumstantial evidence. Defenses, although varied depending on the circumstances, can include the bona fide occupational qualification exception (BFOQ),[12] reasonable factors other than age (RFOA) defense,[13] and bona fide seniority system defense.[14] Although safety professionals will seldom be involved in the actual charge or litigation, it is important for safety professionals to understand the underlying basis for any alleged charge of discrimination based on an employee or applicant's age.

With an expanding population of workers over the age of 40 years entering or currently employed in the American workplace, safety professionals should be cognizant of the protections afforded by the ADEA and OWBPA, as well as individual state laws. Although the scope of the ADEA is substantially narrower than Title VII, prudent safety professionals should take note of their actions and inactions that may impact, directly or indirectly, the older workers and applicants over the age of 40 years. Safety professionals should strive to ensure the reasonableness and practicability of their decision making in light of the protections afforded to individuals under the ADEA, OWBPA, and state statutes. Safety professionals should always strive to maintain fair employment practices for all individuals derived through a fair, equitable, and ethical manner that affords legal protections to the company or organization in which they are employed.

CHAPTER QUESTIONS (TRUE/FALSE)

1. The OWBPA prohibits older workers from participating in a 401K plan.
2. ADEA protects workers above the age of 30.
3. ADEA is enforced by the EEOC.
4. An ADEA charge must be filed within 3 years of the date of incident.
5. States may enact laws providing greater protection than ADEA.

ANSWERS

1. False
2. False
3. True
4. False
5. True

ENDNOTES

1. Cavico, F. J. and Mujtaba, B. G. Discrimination and the Aging American Workforce: Legal Analysis and Management Strategies, *Nova Southwestern University, Journal of Legal Issues and Cases in Business*, 1, 2009.
2. Cavico, F. J., and Mujtaba, B. G., *Journal of Legal Issues and Cases in Business*, p. 2.
3. 29 USCA Section 621 et. seq.
4. Ibid.

5. 29 USCA Section 623(d).
6. 29 USCA Section 623(c).
7. *Allen v. Marshall Fields & Co.*, 93 FRD 438 (ND Ill., 1982).
8. Older Workers Benefit Protection Act (2008), Public Law 101-433; also see EEOC website: www.eeoc.gov.
9. 29 USCA Section 626(d).
10. 29 USCA Section 1627.3(b)(2); also see 29 CFR 1627.3(b)(3) and 1627.5(a).
11. 29 USCA Section 255(a).
12. 29 USCA Section 623(f).
13. 29 CFR Section 860.103(c).
14. 29 USCA Section 623(f)(2).

9 Wage and Hour Laws

If you divorce capital from labor, capital is hoarded, and labor starves.

—Daniel Webster

The surest way to establish your credit is to work yourself into the position of not needing any.

—Maurice Switzer

Learning objectives:

1. Acquire an understanding of the Fair Labor Standards Act requirements.
2. Acquire an understanding of the exempt category of employment.
3. Acquire an understanding of the concept of working time.
4. Acquire an understanding of the child labor laws.
5. Acquire an understanding of the Equal Pay Act.
6. Acquire an understanding of the Lilly Ledbetter Fair Pay Act.

Safety professionals should be aware that although the issues of the payment of wages and the hours worked can appear simplistic from afar, the myriad laws and regulations involved in the spectrum of issues having to do with wages and hours worked are very complex. Safety professionals are usually not directly involved in the day-to-day management of the wages of employees, hours worked, and related issues; however, it is important that safety professionals possess a firm grasp of the requirements of each of the various laws and regulations impacting this area in order to avoid the potential of costly errors in the areas where these laws and regulations may impact the safety function.

The earning of wages or monies paid for the labor provided is one of the primary reasons why your employees show up for work. The U.S. Department of Labor is the primary agency that governs the minimum wages to be paid, as well as the payment to be provided for overtime, as proscribed under the Fair Labor Standards Act (FLSA). Within the structure of the U.S. Department of Labor is the Wage and Hour Division, which specifically addresses the standards established by the FLSA governing the areas of minimum wages, overtime pay, recordkeeping, and child labor. In general, the FLSA applies to companies and organizations functioning within interstate commerce, generating $500,000 in annual dollar volume, and specific sectors such as hospitals, schools, and governmental agencies. Employees covered under the minimum wage and overtime provisions of the FLSA are often called nonexempt employees.

Safety professionals, being part of the management team, should be aware that the safety function is often within the exempt category for both minimum wage and overtime. Certain categories of employees, the largest of which is executive, administrative, and professional employees, are not required to comply with the minimum wage and overtime provisions of the FLSA. Virtually all other employees in most companies or organizations are covered within the FLSA, and thus the minimum wage and overtime provisions are applicable.

Effective July 24, 2009, the FLSA requires companies and organizations to provide nonexempt employees a minimum wage of not less than $7.25 per hour. Safety professionals should be aware that there are exceptions within the minimum wage requirements, such as youths under the age of 20, where the minimum wage is $4.25, employees paid on a piece-rate basis, and other jobs, such as waiters, where gratuities or tipping is involved. Safety professionals should also be cognizant that the FLSA establishes only the minimum wage to be provided; it does not address the maximum wage in any way. Additionally, individual states have also enacted laws in this area, providing a higher minimum wage within their state. (Note: Please see Appendix 3 for differences in each of the individual state's minimum wage.)

Safety professionals should be aware of what qualifies as working time, and thus payment of wages is required. In general, working time includes all of the time when the employee is on duty on the company or organization's worksite or at a particular work location. Under the portal-to-portal exclusion, activities occurring before or after the end of a work shift, such as clothes changing time, are an issue that safety professionals should check with their human resources department about. Safety professionals should be aware that working time can include such activities as rest breaks or meal breaks, off-duty waiting time, on-call waiting time, sleeping time, and training time. Activities that usually constitute noncompensable time include employment application and testing, medical attention outside of working hours, bona fide meal or rest periods, vacations, holidays, and travel to and from work. Safety professionals should also be aware that the FLSA does permit the reasonable cost or fair value of noncash items to be included in wages for the purpose of satisfying the minimum wage requirements. Some of these noncash items can include the fair value for meals, merchandise, tuition, union dues, and insurance premiums.

Although often confusing, safety professionals should be aware that the FLSA does not limit the number of hours in any given day or any given week in which a company or organization can require an employee to work. However, the FLSA does limit the number of hours for younger workers at least 16 years of age. Additionally, the FLSA does require companies or organizations to pay nonexempt employees not less than one and one-half times the employee's regulate rate of pay for all hours in excess of 40 in any given work week. Safety professionals also should be aware that there are posting and recordkeeping requirements under the FLSA enforced by the Wage and Hour Division; however, the FLSA does not contain any specific reporting requirements. The Wage and Hour Division, similar to OSHA, does conduct on-site inspections, and the records must be maintained by the company or organization and open for inspection.

Another area within the FLSA of importance to safety professionals is that of the child labor rules. Safety professionals should be aware that the rules vary depending on the age of the young individual, his or her intended job function, and whether the job function is agriculture related or nonagricultural. Focusing on nonagricultural work, the child labor protections require proscribed hours of work and occupations for individuals under the age of 16, and specific hazardous occupations that are prohibited for individuals under the age of 17. More specifically, the following are permissible jobs and hours for young individuals under the age of 18, as set forth in 29 CFR Part 570:

- Minors age 18 or older are not subject to restrictions on jobs or hours.
- Minors age 16 and 17 may perform any job not declared hazardous by the secretary of (labor), and are not subject to restriction on hours.
- Minors age 14 and 15 may work outside of school hours in various non-manufacturing, nonmining, nonhazardous jobs listed by the secretary (of labor) in regulations, under the following conditions:
  - No more than 3 hours on a school day
  - Eighteen hours in a school week
  - Eight hours on a nonschool day
  - Forty hours in a nonschool week.[1]
- Minors cannot begin work before 7:00 a.m. or work after 7:00 p.m. (except June 1 through Labor Day, when evening hours extend to 9:00 p.m.).
- Permissible work for 14- and 15-year-olds is limited to jobs in retail, food service, and gasoline service establishments specifically listed in the regulations.
- Work Experience and Career Exploration program students may work up to 23 hours during school weeks and 3 hours on school days (including during school hours).[2]

Safety professionals should be cognizant of the overtime provisions, especially when conducting safety training, as well as the age requirements in such areas as the use of summer part-time workers.

Safety professionals should be aware that in addition to the FLSA, several other federal laws, including the Sugar Act,[3] Walsh-Healey Act,[4] and Interstate Commerce Act,[5] specifically address child labor issues, and numerous states address the employment of children in certain industries with more stringent laws. Prudent safety professionals should check with their human resources departments regarding specifics in the areas of overtime, minimum wages, and child labor.

Within the wage and hour area, safety professionals should be aware of the Equal Pay Act of 1963, which is an adjunct to the FLSA.[6] This law addresses discrimination in wage rates based on the sex of male and female employees doing jobs that require equal skills, efforts, and responsibilities and performed under similar working conditions, unless the company or organization can justify the difference through one of the exceptions. Safety professionals should be aware that in 1978, the administration and enforcement of the Equal Pay Act was transferred from the U.S. Department of Labor to the Equal Employment Opportunity Commission.[7]

Safety professionals should be aware that although the term *equal* is used in the Equal Pay Act, the courts have interpreted this term to mean "substantially equal" and not identical.[8] Specifically, the courts have evaluated the four areas or standards of skill, effort, responsibility, and working conditions to determine whether or not a company or organization is in compliance with this law. Safety professionals should also note that the Equal Pay Act only applies to wage discrimination and does not extend to wage-influencing factors, such as incentive programs, merit programs, seniority programs, training programs, and related programs. Lastly, safety professionals should be aware that the Equal Pay Act does possess recordkeeping requirements, which are usually managed by most companies' human resources departments.

Safety professionals should be aware of the relatively new Lilly Ledbetter Fair Pay Act, which was enacted in 2009. This law amended Title VII, ADEA, ADA, and the Rehabilitation Act to clarify that the discriminatory act occurred each and every time compensation was paid in violation of the law. This law was passed in response to a U.S. Supreme Court decision in *Ledbetter v. Goodyear Tire & Rubber Co.*,[9] where the court held that an individual must file an EEOC charge within 180 days of the alleged unlawful employment practice occurring. The Lilly Ledbetter Fair Pay Act changed this decision, permitting workers who do not realize that they have been subjected to discriminatory practices until months or years after the discrimination occurred can extend the time limits so that the event is triggered each time the employee is affected by the unlawful practice. This law permits back pay with a timely filed charge with the EEOC (180 day federal limit and usually 300 days under state law). However, relief is limited to the 2 years preceding the filing of the charge.

Within the area of wages and hours, safety professionals should be aware of the Consumer Credit Protection Act (CCPA),[10] which was designed to protect borrowers of money, requiring the full disclosure of financing terms and conditions. However, housed in Title VI of this law is the Fair Credit Reporting Act, which applies to companies and organizations in the acquisition of credit information regarding employees or applicants. Additionally, safety professionals should be aware that this law also addresses the issue of garnishments from employees' earning in any given week. The CCPA also provides protections for employees against termination of their employment as a result of garnishments. Again, as above, enforcement of the CCPA is by the Wage and Hour Division of the U.S. Department of Labor.

Although safety professionals seldom work directly in the area of wages and hours, the impact of these laws can be felt throughout the safety function. Safety professionals should be cognizant of these laws and their impact and ensure that compliance is maintained in all activities and functions. If there is ever a doubt in the safety professional's mind as to the requirements of these laws, he or she should talk with his or her human resources department before proceeding with any activity, discussion, or event that may result in additional costs, such as overtime, or potential violation of these laws and standards. Wage and hour laws on the federal as well as the state level are often taken for granted, however, and can create substantial difficulties for the safety professional if not strictly adhered to at all times.

CHAPTER QUESTIONS (TRUE/FALSE)

1. The FLSA governs the issue of overtime pay.
2. Exempt employees qualify for overtime pay.
3. Minimum wage is now $10.00 per hour.

4. The Equal Pay Act required all employees to receive the same pay.
5. The CCPA governs the child labor area.

ANSWERS

1. True
2. False
3. False
4. False
5. False

ENDNOTES

1. 29 USCA Section 201, et. seq.
2. U.S. Department of Labor website: www.dol.gov.
3. 7 USCA Section 304, et. seq.
4. 41 USCA Section 35, et. seq.
5. 49 USCA Section 304, et. seq.
6. 29 USCA Section 206(d).
7. Reorganization Plan No. 1 of 1978.
8. *Schultz v. Wheaton Glass Co.*, 421 F.2d 259 (CA-3, 1970).
9. 550 U.S. 618 (2007).
10. 15 USCA Section 1601 et. seq. (1972).

10 Federal Retirement and Welfare Laws

The wagon rests in winter, the sleigh in summer, the horse never.

—**Yiddish Proverb**

Rest and motion, unrelieved and unchecked, are equally destructive.

—**Benjamin Cardozo**

Learning objectives:

1. Acquire a working knowledge of the Employment Retirement Income Security Act.
2. Acquire an understanding of the requirements of the Comprehensive Omnibus Budget Reconciliation Act.
3. Acquire an understanding of the requirement of the Federal Employees' Compensation Act.
4. Acquire a basic understanding of the concepts used within state workers' compensation programs.

With the aging American workforce, safety professionals should be cognizant of the various laws that impact retirement, insurance, protection of medical information, and related areas. Safety professionals often encounter issues that are impacted by these laws and, in many cases, have direct responsibility for such areas as state workers' compensation administration, management of medical records, and management of health insurance programs.

The Employment Retirement Income Security Act (ERISA)[1] was primarily enacted to protect individuals in employee benefit programs through establishing disclosure and reporting requirements, establishing fiduciary standards of conduct, creating penalties, and providing employees access to the federal courts against their companies or organizations and others. ERISA is divided into four different titles involving three different federal agencies.

Title I of ERISA primarily addresses employee rights and the protections afforded to employees. The administering agency is the U.S. Department of Labor. Title II is primarily amendments to the Internal Revenue Code and is administered by the Internal Revenue Service. Title III identifies the jurisdiction of ERISA as well as administration and enforcement. The administration of Title III is shared by the U.S. Department of Labor, Internal Revenue Service, and Pension Benefit Guarantee Corporation. Title IV actually creates the Pension Benefit Guarantee Corporation as well as establishes insurance for employee plan termination and multiemployer retirement plans.

ERISA applies to two general types of employee benefit plans: pension plans and welfare plans (such as medical insurance, disability insurance, etc.). ERISA provides protection against the assignment of employee pension plans, as well as affording protections against retaliation, discrimination, or discharge to the employee's covered plan participant. ERISA also establishes guidelines and requirements to protect employee pension and welfare plans from financial losses caused by mismanagement, misuse, or abuse of the assets within the pension or welfare plan. Safety professionals should also be aware that the reporting and recordkeeping requirements under ERISA are more complex and comprehensive than in virtually any other area of the workplace law arena.

In 1986, Congress passed the Consolidated Omnibus Budget Reconciliation Act (COBRA), which amended ERISA as well as the Internal Revenue Code and Public Health Service Act, to provide temporary continuation of group health benefits to employees who may lose their medical coverage for certain events, such as involuntary termination, layoffs, or reduction in hours. Of importance to safety professionals, COBRA applies to private sector companies with twenty or more employees with a health plan; the company or organization is required to notify the plan administrator within 30 days of the qualifying event (e.g., death, termination, reduction in hours, or entitlement to Medicare). Usually the plan administrator sends an election notice to the affected employee within 14 days of receipt of notice from the company or organization, and the affected employee then has 60 days to decide whether or not to elect COBRA continuation coverage. Additionally, the affected employee has 45 days after electing coverage to pay the initial payment or premium. Safety professionals should be aware that the affected employee is responsible for the payment of the COBRA continuation insurance coverage, and generally the coverage cannot exceed a maximum of 18 months. Just like other insurances, COBRA coverage begins when the group coverage is lost due to the qualifying event and stops if the premium is not paid, the company stops the group health plan, or another event specified in the law.

In general, the affected individual who qualified for COBRA coverage usually pays the entire group health premium amount, including the amount the individual usually paid as an employee and the amount paid by the company or organization. The plan administrator may also charge a 2% administration fee. However, safety professionals should be aware that under the American Recovery and Reinvestment Act of 2009 (ARRA), qualified individuals and family members eligible as a result of an involuntary termination from September 1, 2008, through December 31, 2009, who applied for COBRA coverage may be eligible to pay a reduced premium amount, only 35% of the premium cost, for up to 9 months.

Safety professionals should be aware that COBRA is administered by several federal agencies. The U.S. Department of Labor and Internal Revenue Service–Treasury Department have jurisdiction over private sector group health plans, and the Department of Health and Human Services governs public sector health plans. Safety professionals should be aware that the U.S. Department of Labor provides the regulations and interpretations addressing disclosure and notification requirements, while the Internal Revenue Service–Department of the Treasury issues regulations regarding eligibility, coverage, and premiums.

Although safety professionals may think of the Social Security Act as a retirement benefit, it is important that safety professionals also recognize that the Social Security Act also provides protection in cases of disability and death.[2] Social security benefits are derived primarily from the employment relationship. For most, Social Security benefits are derived primarily from through employment relationship, with employees paying into the system while working, and benefits being paid to employees who become disabled or who reach the proscribed age for retirement. Individuals who have worked and fulfilled the age requirement established by statute are considered fully insured and can acquire a designated insurance amount. Disability benefits are provided to individuals under the age of 65 who are unable to engage in any substantial gainful activity because of any medically determined physical or mental impairment that can be expected to result in death or that has lasted or can be expected to last for a continuous period of not less than 1 year.[3]

Safety professionals often encounter the Social Security Act when employees incur non-work-related debilitating injuries or illnesses or an employee is killed off of the job. Safety professionals should be aware that there is a formal application process to the Social Security Administration, and social security benefits are not automatically paid to impaired individuals. Safety professionals should be aware that if the claim is denied by the Social Security Administration, there is an extensive administrative appeals process established with the Social Security Act, as well as potential appeal to the federal court system after all administrative appeals are exhausted.

Another area that may impact the safety function within the Social Security Act is unemployment insurance. This nationwide joint federal and state system provides unemployment benefits to protect individuals and families against the loss of income due to unemployment.[4] This complex system is encompassed within the Social Security Act as well as the Federal Unemployment Tax Act,[5] requiring employers to provide an unemployment tax to the state at varying rates for employees, and the state to provide the funds to the U.S. Treasury Department. The secretary of the treasury is responsible for investing the funds, and the U.S. Department of Labor is responsible for certifying the unemployment payments to the state agencies. Each state has individual unemployment compensation laws that incorporate various provisions of the Social Security Act and Federal Unemployment Tax Act.

The Federal Employees' Compensation Act provides workers' compensation coverage to civilian employees who become disabled or die while employed with the federal government. Most safety professionals working in the private sector are more familiar with their individual state workers' compensation laws and regulations, which are on a state-by-state basis. In general, many safety professionals have found that the management and administration of their company's workers' compensation program is often conjoined with the safety function. Safety professionals should possess a firm grasp of the basic structure, mechanics, and laws related to their state's workers' compensation system, as well as the specific rules, regulations, and requirements of their state system.

In general, virtually all workers' compensations systems are fundamentally no-fault mechanisms through which employees who incur work-related injuries, illnesses, or death are compensated with monetary and medical benefits. All state workers' compensation laws require an employer-employee relationship for workers' compensation coverage. In a nutshell, the concept of workers' compensation is a compromise in that employees are guaranteed a percentage of their wages and full payment for their medical costs when injured on the job. Employers, on the other hand, are guaranteed a reduced monetary cost for these injuries or illnesses and are provided a protection from additional or future legal action by the employee for the injury.

In general, most workers' compensation systems possess the following features:

- Every state in the United States has a workers' compensation system. There may be variations in the amounts of benefits, the rules, administration, etc., from state to state. In most states, workers' compensation is the exclusive remedy for on-the-job injuries and illnesses.
- Coverage for workers' compensation is limited to employees who are injured on the job. The specific locations as to what constitutes the work premises and on the job may vary from state to state.
- Negligence or fault by either party is largely inconsequential. No matter whether the employer is at fault or the employee is negligent, the injured employee generally receives workers' compensation coverage for any injury or illness incurred on the job.
- Workers' compensation coverage is automatic; e.g., employees are not required to sign up for workers' compensation coverage. By law, employers are required to obtain and carry workers' compensation insurance or be self-insured.
- Employee injuries or illnesses that "arise out of and in the course of employment" are usually considered compensable. These definition phrases have expanded such injuries and illnesses beyond the four corners of the workplace, to include work-related injuries and illnesses incurred on the highways, at various in- and out-of-town locations, and other such remote locales. These two concepts, "arising out of" the employment and "in the course of" the employment, are the basic burdens of proof for the injured employee. Most states require both. The safety and health professional is strongly advised to review the case law in his or her state to see the expansive scope of these two phrases. That is, the injury or illness must "arise out of"; i.e., there must be a causal connection between the work and the injury or illness, and it must be "in the course of" the employment; this relates to the time, place, and circumstances of the accident in relation to the employment (see selected case summary). The key issue is a work connection between the employment and the injury or illness.[6]
- Most workers' compensation systems include wage-loss benefits (sometimes known as time-loss benefits), which are usually between one-half and three-fourths of the employee's average weekly wage. These benefits are normally tax-free and are commonly called temporary total disability (TTD) benefits.

- Most workers' compensation systems require payment of all medical expenses, including hospital expenses, rehabilitation expenses, and prosthesis expenses.
- In situations where an employee is killed on the job, workers' compensation benefits for burial expenses and future wage-loss benefits are usually paid to the dependents.
- When an employee incurs an injury or illness that is considered permanent in nature, most workers' compensation systems provide a dollar value for the percentage of loss to the injured employee. This is normally known as permanent partial disability (PPD) or permanent total disability (PTD).
- In accepting workers' compensation benefits, the injured employee is normally required to waive any common law action to sue the employer for damages from the injury or illness.
- If the employee is injured by a third party, the employer usually is required to provide workers' compensation coverage, but can be reimbursed for these costs from any settlement that the injured employee receives through legal action or other methods.
- Administration of the workers' compensation system in each state is normally assigned to a commission or board. The commission or board generally oversees an administrative agency located within state government that manages the workers' compensation program within the state.
- The Workers' Compensation Act in each state is a statutory enactment that can be amended by the state legislatures. Budgetary requirements are normally authorized and approved by the legislatures in each state.
- The workers' compensation commission or board in each state normally develops administrative rules and regulations for the administration of workers' compensation claims in the state.
- In most states, employers with one or more employees are normally required to possess workers' compensation coverage. Employers are generally allowed several avenues to acquire this coverage. Employers can select to acquire workers' compensation coverage from private insurance companies, from state-funded insurance programs, or become self-insured.
- Most state workers' compensation coverage provides a relatively long statute of limitations. For injury claims, most states grant between 1 and 10 years to file the claim for benefits. For work-related illnesses, the statute of limitations may be as high as 20 to 30 years from the time the employee first noticed the illness or the illness was diagnosed. An employee who incurred a work-related injury or illness is normally not required to be employed with the employer when the claim for benefits is filed.
- Workers' compensation benefits are generally separate from the employment status of the injured employee. Injured employees may continue to maintain workers' compensation benefits even if the employment relationship is terminated.
- Most state workers' compensation systems possess some type of administrative hearing and appeal process.[7]

Safety professionals should be aware of the "graying" of the American workforce and the impact of this aging on not only the safety function but also the laws governing this employee group. "As of the turn of the century, there were about 7 elderly persons for every 100 persons 18 to 64 years. By 1982, that ratio was almost 19 elderly persons per 100 persons of working age. By 2000, that ratio is expected to increase to 21 per 100 and then surge to 38 per 100 by 2050."[8] Safety professionals should be aware of the various laws that directly or indirectly impact the safety function, as well as the employees and the protections and benefits earned through their work activities.

CHAPTER QUESTIONS (TRUE/FALSE)

1. COBRA is designed to address workers' compensation issues.
2. ERISA stands for Early Retirement Insurance Safety Act.
3. Social Security only provides retirement benefits.
4. Workers' compensation is for work-related injuries or illnesses only.
5. The American workforce is getting younger.

ANSWERS

1. False
2. False
3. False
4. True
5. False

ENDNOTES

1. 29 USCA Section 1001, et. seq.
2. 72 USCA Section 501, et. seq.
3. 26 USCA Section 3301.
4. 5 USCA Section 8503.
5. 26 USCA Section 3301.
6. See generally, Larson, L. K. and Larson, A., *Workers Compensation Law: Cases, Materials and Text*, 3rd ed., Matthew Bender, 2000.
7. Schneid, T. D., *Legal Liabilities for Safety Professionals*, 2010.
8. U.S. Senate Special Committee on Aging, *Aging America: Trends and Prospects*, 1981.

11 Privacy Laws

I am dying with the help of too many physicians.

—**Alexander the Great**

Medicine is the only profession that labors incessantly to destroy the reason for its existence.

—**James Bryce**

Learning objectives:

1. Acquire an understanding of the requirements of the Privacy Act of 1974.
2. Acquire an understanding of the sunshine and open meeting laws.
3. Acquire a working knowledge of the Freedom of Information Act.
4. Acquire an understanding of the Health Insurance Portability and Accountability Act.
5. Acquire an understanding of the Genetic Information Nondiscrimination Act.
6. Acquire an understanding of the Fair Credit Reporting Act.

Safety professionals should be aware that there are a number of laws safeguarding individuals' rights to privacy. These laws, on both a federal and a state level, encompass a broad spectrum of privacy issues, ranging from wiretapping to whistleblowing, and offer recourse to individuals whose rights have been violated. Although most safety professionals will have limited direct interaction with these laws, it is important that a prudent safety professional be aware of these laws so not to inadvertently violate any individual or employee's rights while performing the duties and responsibilities of the safety function.

The Privacy Act of 1974[1] generally requires federal agencies to maintain records of individuals in a current, accurate, and relevant manner, and to permit the individual to gain access to any records pertaining to him or her. Additionally, the federal agency is prohibited from refusing to disclose the records or information about the individual, except in very limited circumstances, and provides a method through which individuals can make corrections to the federal agency records.

Safety professionals should be aware that there is a proscribed procedure for the individual to appeal any refusal by the federal agency in correcting the record. Specific monetary penalties are provided in this law if the federal agency acts in an intentional or willful manner to fail to maintain accurate records, and individual employees of the federal agency can be penalized for willfully disclosing prohibited records. Safety professionals should be aware that federal agencies may disclose or release records of individuals, with or without the individual's consent, in specific circumstances, in order to fulfill the requirements of the Freedom of Information

Act.[2] Safety professionals should be aware that "any person who knowingly and willfully requests or obtains any record concerning an individual from the federal agency under false pretenses shall be guilty of a misdemeanor and subject to a fine of not more than $5,000.00."[3]

Correlating with the Privacy Act is the Sunshine Act, which requires federal agencies to open their meetings to the public. This federal law often is paralleled by state open meeting laws. The purpose of this law is to provide the public with the "fullest practicable information regarding the decision-making processes of the federal government."[4] Safety professionals should be aware that certain federal agencies, such as law enforcement, are exempt from the open meetings requirement.

Of particular importance to safety professionals acquiring information from OSHA or other federal governmental agencies is the Freedom of Information Act[5] (FOIA). This law requires federal agencies to make its rules, opinions, orders, records, and proceedings available to the general public, as well as requiring the federal agency to publish information, such as new OSHA standards, in the *Federal Register*, which provides guidance to employers and employees, as well as the public in general.

More specifically, the FOIA requires each federal agency to publish the following in the *Federal Register*:

- Descriptions of its central and field organizations and the established places at which, the employees from whom, and the methods whereby, the public may obtain information, make submittals or requests, or obtain decisions;
- Information concerning the general course and method by which its functions are channeled and determined, including the nature and requirements of all formal and informal procedures available;
- Procedural rules, descriptions of forms available or the places at which forms may be obtained, and instructions as to the scope and contents of all papers, reports, or examinations;
- Substantive rules of general applicability adopted as authorized by law, and related statements or interpretations of general policies;
- Amendments, revisions, or repeals of the above.[6]

FOIA is important for safety professionals in that this statute requires OSHA and other governmental agencies to publish their standards and procedures in the *Federal Register*. Prudent safety professionals may want to monitor the *Federal Register* for proposed or new standards or changes in policy or procedures. Although OSHA also publishes its opinions on its website, safety professionals should be aware that federal governmental agencies are required to also make available for review and copying other information, such as final orders, adjudication of cases, and interpretations and statements of policy that are not required to be published in the *Federal Register*. Additionally, safety professionals should be aware that such materials as field manuals, instruction materials, and related manuals can be published and copies offered for sale by the agency,[7] and the agency may charge for copying the requested documents. Safety professionals should be aware that personnel and medical records are exempt from this law.[8]

In the area of medical records, safety professionals should be aware of the Health Insurance Portability and Accountability Act (HIPAA)[9] requirements, especially focusing on the security requirements for medical information. HIPAA was enacted in 1996, requiring the Department of Health and Human Services to adopt a national standard for electronic health care transactions and healthcare information security. In 2000, the Department of Health and Human Services published the final privacy rules, which were amended in 2002, establishing national standards for the protection of individually identifiable health information for health plans, health care clearinghouses, and health care providers who utilize electronic transfer of medical information. In 2003, the Department of Health and Human Services published the final security rule establishing national standards for protecting the confidentiality, integrity, and availability of electronic protected health information.[10] Safety professionals managing an on-site medical facility, the workers' compensation function, and related functions requiring access to individual employee medical records should pay careful attention to the requirements of HIPAA.

Correlating to HIPAA, safety professionals should be aware of the Genetic Information Nondiscrimination Act of 2008 (GINA).[11] Safety professionals should be aware that Title I of GINA prohibits group health plans and health insurance issuers from discriminating against individuals based upon genetic information. Safety professionals should be aware that *genetic information* "means information about an individual's genetic tests, the genetic tests of family members of the individual, the manifestation of a disease or disorder in family members of the individual or any request for or receipt of genetic services, or participation in clinical research that includes genetic services by the individual or family member of the individual. The term genetic information includes, with respect to a pregnant woman (or a family member of a pregnant woman) genetic information about the fetus and with respect to an individual using assisted reproductive technology, genetic information about the embryo."[12]

Safety professionals should be aware that *genetic services* means "genetic tests, genetic counseling or genetic education," and *genetic testing* means "an analysis of human DNA, RNA, chromosomes, proteins or metabolites."[13] Safety professionals should be aware that GINA prohibits group health plans from requesting or requiring individuals from undergoing genetic testing or collecting genetic information as part of a medical assessment for insurance or underwriting purposes. Although safety professionals usually do not encounter genetic information or include genetic testing in the employment process, safety professionals should be aware of GINA as new and cost-effective technologies emerge in the evaluation of individuals in the workplace.

Within the area of protections provided to individuals' information, safety professionals should be aware of the Fair Credit Reporting Act (FCRA).[14] This law provides protections for applicants and employees, prohibiting companies or organizations from acquiring an investigative consumer report to be used to evaluate an individual for employment purposes. However, safety professionals should be aware that companies or organizations may acquire this financial information if the company or organization mails a written notice to the applicant or employee that a consumer investigative report will be made and the company or organization provides a complete disclosure as to the nature and scope of the investigation.[15] Safety professionals should be aware that a consumer investigative report may contain information about the individual's character, general reputation, personal characteristics, and mode of living, as well as financial information.

Safety professionals should be aware that if a consumer investigative report is utilized and employment is denied by the company or organization, the company or organization is required to advise the applicant or employee of the reason and provide the name and address of the consumer reporting agency that conducted the investigation and generated the report.[16] If the company or organization or the consumer reporting agency willfully or negligently fails to follow the FCRA, the individual is entitled to pursue a civil action in the appropriate U.S. District Court.

Lastly, safety professionals should be aware that numerous laws provide protections to whistleblowers, individuals or employees who report violations of the law to a governmental agency. For safety professionals working in the public sector, the Civil Service Reform Act[17] prohibits retaliation against an applicant or employee for disclosing information "he or she believes to be a violation of any law, rule, regulation, or mismanagement, a gross waste of funds, an abuse of authority, or a substantial danger to public health or safety."[18] Under this law, the Office of Special Counsel is provided the responsibility and duty to investigate claims of reprisals for employment against other federal employees or applicants who report these violations.[19]

Prudent safety professionals should exercise caution at any time individual privacy issues are involved in any situation. Safety professionals should review any privacy issues with their human resources department or legal counsel before taking any action that may violate any of the protections provided to applicants or employees and their personal or medical information.

CHAPTER QUESTIONS (TRUE/FALSE)

1. Any citizen can request information under the FOIA.
2. GINA addresses DNA testing information.
3. HIPAA only addresses internal accounting records.
4. FOIA requires a face-to-face meeting to request documents.
5. The FCRA addresses the requirements of the OSHA LOTO standard.

ANSWERS

1. True
2. True
3. False
4. False
5. False

ENDNOTES

1. 5 USCA Section 552a; 5 CFR Section 297.101.
2. 5 USCA Section 552a(b)(1).

3. 5 USCA Section 552a(i).
4. Ibid.
5. 5 USCA Section 552.
6. 5 USCA Sections 552(a)(1)(A)–(e).
7. 5 USCA Section552(a)(2)(A)–(C).
8. 5 USCA Section (b)(1)–(9).
9. Public Law 104-191 (1996).
10. Department of HHS website: www.hhs.gov.
11. Department of Labor: www.dol.gov.
12. Ibid.
13. Ibid.
14. 15 USCA Section 1681d.
15. Ibid.
16. 15 USCA Section 1681m.
17. 5 USCA Section 2302(b)(8).
18. Ibid.
19. 8 USCA Section 1104, et. seq.

12 Family and Medical Leave Act and Safety

Heredity: the thing a child gets from the other side of the family.

—**Marcelene Cox**

Family is the most effective form of government.

—**Robert Half**

Learning objectives:

1. Acquire a general understanding of the Family and Medical Leave Act.
2. Acquire a general understanding of the National Defense Authorization Act.
3. Acquire a general understanding of the Uniformed Services Employment and Reemployment Rights Act.
4. Acquire an understanding of the interaction of the FMLA and the safety function.

Safety professionals with responsibilities for the management of staff personnel or the management and administration of the workers' compensation function should become well versed in the requirements of the Family and Medical Leave Act (FLMA),[1] as well as the National Defense Authorization Act (NDAA).[2] In general, the FMLA "provides a means for employees to balance their work and family responsibilities by taking unpaid leave for certain reasons ... and is intended to promote the stability and economic security of families as well as the nation's interest in preserving the integrity of families."[3] The NDAA amends the military leave entitlements of the FMLA, expanding coverage to family members, military caregivers, and spouses, sons, daughters, parents, or next of kin for certain veterans with serious injuries or illnesses.[4]

For safety professionals working in the private sector, the basic requirements for an employee to be eligible for a leave of absence of up to 12 work weeks that is unpaid but job protected start with employment. Under the FMLA, the employee must meet the eligibility requirements, which include:

- Be employed by a covered employer and work at a worksite within 75 miles of which that employer employs at least 50 people;
- Have worked at least 12 months (which does not have to be consecutive) for the employer; and
- Have worked at least 1,250 hours during the 12 months immediately before the date FMLA leave begins.[5]

For most safety professionals, the majority of the private sector companies or organizations meet this requirement. However, safety professionals should be aware that some companies and organizations have established policies in which the requirements to qualify for FMLA are substantially lower than the federal minimum requirements. For an employee to be eligible to take a job-protected, unpaid FMLA leave of absence, the employee must identify a qualifying event, which includes:

- Birth and care of the employee's child, within 1 year of birth
- Placement with the employee of a child from adoption or foster care, within 1 year of the placement
- Care of an immediate family member (spouse, child, parent) who has a serious health condition
- A health condition of the employee that makes the employee unable to perform the essential functions of his or her job

Safety professionals should be aware that most private sector companies or organizations have policies and procedures established, usually through the human resources department, to address all requests for FMLA leave of absences. Safety professionals often see requests for FMLA leave for non-work-related injuries or illnesses by employees or family members; however, FMLA leave can also be requested in cases involving work-related injuries or illnesses. Safety professionals should be aware that although the employee is eligible for up to 12 work weeks of unpaid leave, employees may take the leave of absence on an intermittent basis or on a reduced leave schedule if medically necessary or due to a qualifying exigency. Additionally, safety professionals should be aware that intermittent leave for the placement of adoption, foster care of a child, or birth of a child is subject to approval by the company or organization. However, safety professionals should be aware that approval by the company or organization is not required for intermittent schedules or reduced work schedules that are medically necessary due to pregnancy, serious health conditions, or serious illness or injury of a covered service member, or in exigent circumstances.

Although many instances in which an FMLA leave of absence is requested are unforeseen, safety professionals should be aware that if the requested leave of absence is foreseeable, the employee is required to provide the company or organization at least 30 days' notice, or as soon as practicable. The employee is required to comply with the company or organization's policies and procedural requirements for requesting an FMLA leave of absence and provide sufficient information to permit the company or organization to make a reasonable determination of whether FMLA applies to the leave of absence request. Safety professionals should be aware that in situations where the company or organization has previously granted an FMLA leave of absence and the employee is requesting a second FMLA leave, the employee must specifically reference the qualifying reason or need for a second FMLA leave of absence.

Although it is often difficult for an employee to financially support 12 weeks without a paycheck in these difficult financial times, safety professionals should be aware that the company or organization can request that the employee requesting an FMLA leave of absence for a serious health condition provide verification and certification of the employee's health condition or the health condition of the family member or service member. Additionally, safety professionals should be aware that companies or organizations granting an FMLA leave of absence can require a periodic report on the employee's health status or family member or service member's health status. Of importance to safety professionals, the company or organization may require that the employee returning from an FMLA leave of absence due to a serious health condition provide a certification from the employee's physician or health care provider of the employee's ability to return to full or unrestricted work duty.

Although usually addressed by the company or organization's human resources department, safety professionals should be aware that FMLA provides job protections to employees while they are on a qualified leave of absence. However, when an employee returns from an FMLA leave of absence, the employee is entitled to be restored to same or an equivalent job with equivalent pay, benefits, and other terms and conditions of employment. The company or organization is required to return the employee to employment with the same benefits and at the same level as before the leave of absence; however, the employee is not entitled to any additional benefits accrued during the unpaid leave of absence. Safety professionals should also note that employees may not be employed by other companies or organizations while on FMLA leave from their company or organization.

Safety professionals should be aware that the company or organization is required to maintain any group health insurance benefits during the FMLA leave of absence. The company or organization is also required to maintain the group health insurance benefits at the same level and in the same manner if the employee returns to work. Safety professionals should also note that employees can elect or the company or organization can require that the employee use accrued paid vacation, sick pay, personal days, and other accrued leave for the period of the unpaid FMLA leave of absence. For many companies or organizations, the accrued leave is required to be utilized concurrently with the FMLA leave of absence period of time. Prudent safety professionals should check with their human resources department to identify the company or organization's policy and procedure regarding concurrent FMLA and accrued leave.

The governing federal agency for FMLA is the Wage and Hour Division of the U.S. Department of Labor. Safety professionals should be aware that FMLA requires notices and posters to inform employees of their FMLA rights as well as recordkeeping requirements. FMLA does not currently possess any reporting requirements. Safety professionals should note these requirements and may want to incorporate the FMLA poster and notice requirements in their annual audits, wherein the FMLA poster and notice requirements can be verified at the same time as the OSHA posters and other poster requirements.

The National Defense Authorization Act (NDAA)[6] amended the FMLA to allow eligible employees to take up to 12 work weeks of job-protected leave for "qualifying exigency" arising from active duty or call to active duty status of a spouse, son, daughter, or parent. The NDAA also amended the FMLA to permit up to 26 work weeks within a 12-month period to care for a service member with a serious injury or illness. Safety professionals should be aware that these amendments to the FMLA are often referred to as military family leave.[7]

Safety professionals should be aware that the eligibility requirements and employee coverage requirements for NDAA are the same as for FMLA. Under NDAA, "a covered service member is a current member of the Armed Forces, including a member of the National Guard or Reserves, who is undergoing medical treatment, recuperation, or therapy, is otherwise in outpatient status, or is otherwise on the temporary disability retired list, for a serious injury or illness."[8] Safety professionals should be aware that the company or organization is required to provide up to 12 unpaid work weeks of leave during the normal 12-month period for qualifying exigencies arising from the employee's spouse, son, daughter, or parent being called to active military duty. However, safety professionals should know that under the terms of NDAA, qualifying exigency leave is only available to family members in the National Guard or Reserves, and does not extend to family members of military service personnel in the regular Armed Forces.

A qualifying exigency under the NDAA includes:

- Short-notice deployment of a military service member for the 7 days from the date of notification
- Military events, ceremonies, programs, and events sponsored by the military or support organizations
- Certain daycare and related activities
- Making financial or legal arrangements
- Attending counseling
- Up to 5 days rest and recuperation leave during deployment
- Attending postdeployment activities
- Other events the employee and company agree to be considered a qualifying exigency event

Safety professionals should be aware that many of the posting and recordkeeping requirements mirror those of the FMLA, and companies or organizations can require verification from the employee requesting an NDAA leave of absence. Safety professionals should be aware that where the military service member and his or her spouse both work for the company or organization, the spouse is limited to a combined total of 26 work weeks in any 12-month period if the leave is to care for a covered service member with a serious injury or illness, and for the birth and care of a newborn child, adoption, foster care of a child, or care of a parent who has a serious health condition.[9]

Safety professionals should be aware that in the vast majority of FMLA and NDAA situations, companies and employers are more than accommodating to the situation of the employee, and the FMLA or NDAA leave is immediately granted. However, it is important for safety professionals to ensure that all forms and verifications required under their company or organization's policies are completed and all necessary certifications required upon the return of the employee.

Similar to FMLA and NDAA is the Uniformed Service Employment and Reemployment Rights Act[10] (USERRA). The USERRA requires that returning veterans receive all benefits of employment that they would have obtained if they had been continuously employed with the company or organization. Safety professionals should be aware that one of these benefits is the ability to receive an FMLA leave of absence. Thus, if a veteran returned from a tour of duty and is reemployed by your company or organization in the same 40-hour-per-week job as he or she had prior to serving, the employee should receive credit toward the 1,250-hour requirements for the time on active duty, and thus would be eligible for FMLA leave.

Safety professionals should be aware that there has been a substantial amount of litigation since the enactment of the FMLA, addressing many of the major elements within this statute. Safety professionals are encouraged to review the information on the U.S. Department of Labor website, as well as the current cases that address many of the issues involving the FMLA and military leave provisions. Safety professionals should possess a level of understanding of their company or organization's policies and procedures with regards to FMLA and military leave, especially in the areas that impact the safety function, such as medical certifications, overlap of FLMA with workers' compensation or short-term disability leave, training requirements, and poster requirements. As always, prudent safety professionals should review any requests related to FMLA, NDAA, or USERRA with their human resources department or legal counsel before proceeding.

CHAPTER QUESTIONS (TRUE/FALSE)

1. The FMLA only provides leave for work-related injuries.
2. FMLA provides a maximum of 12 weeks of unpaid leave.
3. Only employees qualify for FMLA leave.
4. The EEOC governs the provisions of the FMLA.
5. The NDAA provides up to 26 weeks of unpaid leave.

ANSWERS

1. False
2. True
3. True
4. False
5. True

ENDNOTES

1. 29 USC Section 2601; also see Department of Labor website: www.dol.gov.
2. Public Law 111-84 (2010).
3. Department of Labor website: www.dol.gov.
4. Public Law 110-84.
5. Department of Labor: www.dol.gov—Elaws.
6. Public Law 110-181 (2008).
7. Department of Labor website: www.dol.gov.
8. Ibid.
9. Ibid.
10. 38 USC Section 4301–4333 (1994).

13 Americans with Disabilities Act and Safety

There is no problem of human nature which is insoluble.

—Ralph J. Bunche

The purpose of law is to prevent the strong always having their way.

—Ovid

Learning objectives:

1. Acquire an understanding of the requirements of the Americans with Disabilities Act.
2. Identify the interaction of the ADA and the safety function in private sector organizations.
3. Acquire an understanding of the ADA enforcement processes.

One of the more extensive laws that can have a definite impact on the safety function is the Americans with Disabilities Act (ADA). In a nutshell, the ADA prohibits discriminating against qualified individuals with physical or mental disabilities in all employment settings. For safety professionals, the areas of workers' compensation, restricted duty programs, facility modifications, training, and other safety functions are often where the safety function and the ADA may intersect and create duties and responsibilities for the safety professional as well as potential liabilities for the company or organization.

It is important that safety professionals acquire a working knowledge of the ADA as well as key areas in which the ADA and safety function may intersect in order to be able to recognize when the ADA may be applicable to the situation. Safety professionals should be able to recognize when a potentially qualified individual is requesting an accommodation and appropriately address the situation in order to comply with the ADA. As noted by Mr. David Fram, director of ADA and EEO Services from the National Employment Law Institute, in his 2010 presentation, utilization of the simple question "How can I help you?" as the initial response can open the doors to communications between the safety professional and employee, as well as starting down the right path to ensure compliance with the ADA.[1]

Safety professionals should be aware that although the ADA became law in 1990, the ADA is still being molded through court decisions, agency regulations, and interpretations, and was substantially amended by the ADA Amendments Act of 2008. Safety professionals should be aware that effective January 1, 2009, the new Americans with Disabilities Act Amendments Act of 2008 became effective. This new act makes significant changes to the term *disability*, as defined in the ADA, by rejecting several Supreme Court decisions and portions of previous EEOC's ADA regulations. The new act retains the basic definition of *disability* as defined in the ADA as being an impairment that substantially limits one or more major life activities, a record of such an impairment, or being regarded as possessing an impairment. More specifically, the ADA Act Amendments Act provided the following:

1. The Act requires the EEOC to revise the section of their regulations which defines the term "substantially limits";
2. The Act expands the definition of "major life activities" by including two non-exhaustive lists of which list 1 includes many activities that the EEOC has previously recognized (such as walking) as well as activities that the EEOC previously did not specifically recognize (such as reading, bending and communicating) and the second list includes major bodily functions (such as functions of the immune system, normal cell growth, digestive system, bowel, bladder, neurological, brain, respiratory, circulatory, endocrine and reproductive functions);
3. The Act states that mitigating measures such as "ordinary eyeglasses or contact lenses" shall not be considered in assessing whether an individual possesses a disability;
4. The Act clarified that an impairment that is episodic or in remission is a disability if it would substantially limit a major life activity when active; the Act provides that an individual subjected to an action prohibited by the ADA (such as failure to hire) because of an actual or perceived impairment will meet the "regarded as" definition of disability, unless the impairment is transitory or minor;
5. The Act provides that individuals covered only under the "regarded as" prong of the ADA test are not entitled to reasonable accommodation; and
6. The Act emphasizes that the definition of "disability" should be interpreted broadly.[2]

The initial place to start for safety professionals is with the federal or state agency responsible for administration and enforcement of the law. For the ADA, safety professionals should be aware that the federal agency responsible for administration and enforcement of the ADA is the Equal Employment Opportunity Commission (EEOC), and information can be found on its website: www.eeoc.gov. Prudent safety professionals should monitor the EEOC website as well as recent case decisions to ensure the most current information regarding the ADA. Additionally, safety professionals should also be aware that individual states may also possess laws that parallel or are more stringent than the federal level ADA, and information can usually be found on the individual state agency's website.

Safety professionals should be aware that the ADA is divided into five titles, and all titles possess the potential of substantially impacting the safety function in covered public or private sector organizations. Title I contains the employment provisions that protect all individuals with disabilities who are in the United States, regardless of their national origin or immigration status. Title II prohibits discriminating against qualified individuals with disabilities or excluding them from the services, programs, or activities provided by public entities. Title II contains the transportation provisions of the act. Title III, entitled "Public Accommodations," requires that goods, services, privileges, advantages, and facilities of any public place be offered "in the most integrated setting appropriate to the needs of the individual."[1]

Title IV also covers transportation offered by private entities and addresses telecommunications. Title IV requires that telephone companies provide telecommunication relay services, and that public service television announcements that are produced or funded with federal money include closed captions. Title V includes the miscellaneous provisions. This title notes that the ADA does not limit or invalidate other federal and state laws providing equal or greater protection for the rights of individuals with disabilities, and addresses related insurance, alternate dispute, and congressional coverage issues.

Safety professionals in the private sector should pay careful attention to the scope and potential impact of Title I of the ADA on safety functions. Title I prohibits covered employers from discriminating against a "qualified individual with a disability" with regard to job applications, hiring, advancement, discharge, compensation, training, and other terms, conditions, and privileges of employment.[3]

Section 101(8) defines a "qualified individual with a disability" as any person who, with or without reasonable accommodation, can perform the essential functions of the employment position that such individual holds or desires.... Consideration shall be given to the employer's judgment as to what functions of a job are essential, and if an employer has prepared a written description before advertising or interviewing applicants for the job, this description shall be considered evidence of the essential function of the job.[4] The Equal Employment Opportunity Commission (EEOC) provides additional clarification of this definition: "an individual with a disability who satisfies the requisite skill, experience and educational requirements of the employment position such individual holds or desires, and who, with or without reasonable accommodation, can perform the essential functions of such position."[5]

Congress did not provide a specific list of disabilities covered under the ADA because "of the difficulty of ensuring the comprehensiveness of such a list."[6] Under the ADA, an individual has a disability if he or she:

• Possesses a physical or mental impairment that substantially limits one or more of the major life activities of such individual
• Possesses a record of such an impairment
• Is regarded as having such an impairment[7]

Safety professionals should be aware that the ADA utilizes the broader language of *disability* rather than the term *handicapped* adopted under the Rehabilitation Act. For an individual to be considered "disabled" under the ADA, the physical or mental impairment must limit one or more "major life activities." Under the U.S. Justice

Department's regulation issued for Section 504 of the Rehabilitation Act, "major life activities" are defined as "functions such as caring for one's self, performing manual tasks, walking, seeing, hearing, speaking, breathing, learning and working."[8] Congress clearly intended to have the term *disability* broadly construed. However, this definition does not include simple physical characteristics, nor limitations based on environmental, cultural, or economic disadvantages.[9]

The second prong of this definition is "a record of such an impairment disability." The Senate Report and the House Judiciary Committee Report each stated:

> This provision is included in the definition in part to protect individuals who have recovered from a physical or mental impairment which previously limited them in a major life activity. Discrimination on the basis of such a past impairment would be prohibited under this legislation. Frequently occurring examples of the first group (i.e., those who have a history of an impairment) are people with histories of mental or emotional illness, heart disease or cancer; examples of the second group (i.e., those who have been misclassified as having an impairment) are people who have been misclassified as mentally retarded.[10]

The third prong of the statutory definition of a disability extends coverage to individuals who are "being regarded as having a disability." The ADA has adopted the same "regarded as" test that is used in Section 504 of the Rehabilitation Act:

> "Is regarded as having an impairment" means (A) has a physical or mental impairment that does not substantially limit major life activities but is treated ... as constituting such a limitation; (B) has a physical or mental impairment that substantially limits major life activities only as a result of the attitudes of others toward such impairment; (C) has none of the impairments defined (in the impairment paragraph of the Department of Justice regulations) but is treated ... as having such an impairment.[11]

Safety professionals should be aware that a "qualified individual with a disability" under the ADA is any individual who can perform the essential or vital functions of a particular job with or without the employer accommodating the particular disability. Safety professionals should be aware that companies or organizations are provided the opportunity to determine the "essential functions" of the particular job before offering the position through the development of a written job description. This written job description will be considered evidence to which functions of the particular job are essential and which are peripheral. In deciding the essential functions of a particular position, the EEOC will consider the company or organization's judgment, whether the written job description was developed prior to advertising or beginning the interview process, the amount of time spent performing the job, the past and current experience of the individual to be hired, relevant collective bargaining agreements, and other factors.[12]

The EEOC defines the term *essential function* of a job as meaning "primary job duties that are intrinsic to the employment position the individual holds or desires" and precludes any marginal or peripheral functions that may be incidental to the primary job function.[13] The factors provided by the EEOC in evaluating the essential functions of a particular job include the reason that the position exists, the number of employees available, and the degree of specialization required to perform the job.[14] This determination is especially important to safety professionals who may be required to develop the written job descriptions or to determine the essential functions of a given position.

Safety professionals should recognize that they may be placed in a difficult position when the issue involved is whether or not the qualified individual creates a direct threat to the safety and health of himself of herself or others in the workplace. This issue may require the safety professional to evaluate and render a decision that will impact not only the individual with a disability but also the company or organization. Safety professionals should be aware that the ADA does identify that any individual who poses a direct threat to the health and safety of others that cannot be eliminated by reasonable accommodation may be disqualified from the particular job.[15] The term *direct threat* to others is defined by the EEOC as creating "a significant risk of substantial harm to the health and safety of the individual or others that cannot be eliminated by reasonable accommodation."[16] The determining factors that safety professionals should consider in making this determination include the duration of the risk, the nature and severity of the potential harm, and the likelihood that the potential harm will occur.[17] Safety professionals, when addressing this issue, should also consider the EEOC's interpretive guidelines, which state:

[If] an individual poses a direct threat as a result of a disability, the employer must determine whether a reasonable accommodation would either eliminate the risk or reduce it to an acceptable level. If no accommodation exists that would either eliminate the risk or reduce the risk, the employer may refuse to hire an applicant or may discharge an employee who poses a direct threat.[18]

Safety professionals should note that Title I provides that if a company or organization does not make the reasonable accommodations for the known limitations of a qualified individual with disabilities, this action or inaction is to be considered discrimination. However, if the company or organization can prove that providing the accommodation would place an undue hardship on the operation of the business, the claim of discrimination can be disproved. Section 101(9) defines *reasonable accommodation* as:

(a) making existing facilities used by employees readily accessible to and usable by the qualified individual with a disability and includes:
(b) job restriction, part-time or modified work schedules, reassignment to a vacant position, acquisition or modification of equipment or devices, appropriate adjustments or modification of examinations, training materials, or policies, the provisions of qualified readers or interpreters and other similar accommodations for ... the QID (qualified individual with a disability).[19]

The EEOC further defines *reasonable accommodation* as:

1. Any modification or adjustment to a job application process that enables a qualified individual with a disability to be considered for the position such

qualified individual with a disability desires, and which will not impose an undue hardship on the ... business; or

2. Any modification or adjustment to the work environment, or to the manner or circumstances which the position held or desired is customarily performed, that enables the qualified individual with a disability to perform the essential functions of that position and which will not impose an undue hardship on the ... business; or

3. Any modification or adjustment that enables the qualified individual with a disability to enjoy the same benefits and privileges of employment that other employees enjoy and does not impose an undue hardship on the ... business.[20]

Safety professionals should be aware that the company or organization would be required to make reasonable accommodations for any/all known physical or mental limitations of the qualified individual with a disability, unless the employer can demonstrate that the accommodations would impose an undue hardship on the business, or that the particular disability directly affects the safety and health of that individual or others. Safety professionals should also be aware that included under this section is the prohibition against the use of qualification standards, employment tests, and other selection criteria that can be used to screen out individuals with disabilities, unless the employer can demonstrate that the procedure is directly related to the job function. In addition to the modifications to facilities, work schedules, equipment, and training programs, the company or organization is required to initiate an "informal interactive (communication) process" with the qualified individual to promote voluntary disclosure of his or her specific limitations and restrictions to enable the employer to make appropriate accommodations that will compensate for the limitation.[21]

Safety professionals should pay careful attention to Title I, Section 102(c)(1). This section prohibits discrimination through medical screening, employment inquiries, and similar scrutiny. Safety professionals should be aware that underlying this section was Congress's conclusion that information obtained from employment applications and interviews "was often used to exclude individuals with disabilities—particularly those with so-called hidden disabilities such as epilepsy, diabetes, emotional illness, heart disease and cancer—before their ability to perform the job was even evaluated."[22]

Additionally, under Title I, Section 102(c)(2), safety professionals should be aware that conducting preemployment physical examinations of applicants and asking prospective employees if they are qualified individuals with disabilities is prohibited. Employers are further prohibited from inquiring as to the nature or severity of the disability, even if the disability is visible or obvious. Safety and loss prevention professionals should be aware that individuals may ask whether any candidates for transfer or promotion who have a known disability can perform the required tasks of the new position if the tasks are job related and consistent with business necessity. An employer is also permitted to inquire about the applicant's ability to perform the essential job functions prior to employment. The employer should use the written job description as evidence of the essential functions of the position.[23]

Safety professionals may require medical examinations of employees only if the medical examination is specifically job related and is consistent with business necessity. Medical examinations are permitted only after the applicant with a disability has been offered the job position. The medical examination may be given before the applicant starts the particular job, and the job offer may be contingent upon the results of the medical examination if all employees are subject to the medical examinations and information obtained from the medical examination is maintained in separate, confidential medical files. Companies or organizations are permitted to conduct voluntary medical examinations for current employees as part of an ongoing medical health program, but again, the medical files must be maintained separately and in a confidential manner. The ADA does not prohibit safety professionals or their medical staff from making inquiries or requiring medical or fit-for-duty examinations when there is a need to determine whether or not an employee is still able to perform the essential functions of the job, or where periodic physical examinations are required by medical standards or federal, state, or local law.[24]

Correlating with medical testing, safety professionals should pay careful attention to the area of controlled substance testing. Under the ADA, the company or organization is permitted to test job applicants for alcohol and controlled substances prior to an offer of employment under Section 104(d). The testing procedure for alcohol and illegal drug use is not considered a medical examination as defined under the ADA. Companies or organizations may additionally prohibit the use of alcohol and illegal drugs in the workplace and may require that employees not be under the influence while on the job. Companies and organizations are permitted to test current employees for alcohol and controlled substance use in the workplace to the limits permitted by current federal and state law. The ADA requires all employers to conform to the requirements of the Drug-Free Workplace Act of 1988. Thus, safety professionals should be aware that most existing preemployment and postemployment alcohol and controlled substance programs that are not part of the preemployment medical examination or ongoing medical screening program will be permitted in their current form.[25] Individual employees who choose to use alcohol and illegal drugs are afforded no protection under the ADA. However, employees who have successfully completed a supervised rehabilitation program and are no longer using or addicted are offered the protection of a qualified individual with a disability under the ADA.[26]

Safety professionals should be aware that Title III also requires that auxiliary aids and services be provided for the qualified individual with a disability, including, but not limited to, interpreters, readers, amplifiers, and other devices (not limited or specified under the ADA), to provide that individual with an equal opportunity for employment, promotion, etc.[27] Congress did, however, provide that auxiliary aids and services do not need to be offered to customers, clients, and other members of the public if the auxiliary aid or service creates an undue hardship on the business. Safety professionals may want to consider alternative methods of accommodating the qualified individual with a disability. This section also addresses the modification of existing facilities to provide access to the individual, and requires that all new facilities be readily accessible and usable by the individual.

Safety professionals should be aware that Title V identified that the ADA does not limit or invalidate other federal or state laws that provide equal or greater protection for the rights of individuals with disabilities. Safety professionals should also be aware of any individual state laws addressing disability or handicap discrimination in the workplace that may be more restrictive than the ADA.

Safety professionals should be aware that the ADA is substantially broad in nature and provides protections to all individuals associated with or having a relationship to the qualified individual with a disability. This inclusion is unlimited in nature, including family members, individuals living together, and an unspecified number of others. The ADA extends coverage to all individuals, legal or illegal, documented or undocumented, living within the boundaries of the United States, regardless of their status.[28] Under Section 102(b)(4), unlawful discrimination includes "excluding or otherwise denying equal jobs or benefits to a qualified individual because of the known disability of the individual with whom the qualified individual is known to have a relationship or association." Therefore, the protections afforded under this section are not limited to only familial relationships. There appears to be no limits regarding the kinds of relationships or associations that are afforded protection. Of particular note is the inclusion of unmarried partners of persons with AIDS or other qualified disabilities.[29]

As with the OSH Act, the ADA requires that employers post notices of the pertinent provisions of the ADA in an accessible format in a conspicuous location within the employer's facilities. A prudent safety professional may wish to consider providing additional notification on job applications and other pertinent documents.

Under the ADA, safety professionals should be aware that it is unlawful for an employer to "discriminate on the basis of disability against a qualified individual with a disability" in all areas, including:

- Recruitment, advertising, and job application procedures
- Hiring, upgrading, promoting, awarding tenure, demotion, transfer, layoff, termination, the right to return from layoff, and rehiring
- Rate of pay or other forms of compensation and changes in compensation
- Job assignments, job classifications, organization structures, position descriptions, lines of progression, and seniority lists
- Leaves of absence, sick leave, or other leaves
- Fringe benefits available by virtue of employment, whether or not administered by the employer
- Selection and financial support for training, including apprenticeships, professional meetings, conferences and other related activities, and selection for leave of absence to pursue training
- Activities sponsored by the employer, including social and recreational programs
- Any other term, condition, or privilege of employment[30]

Safety professionals should be aware that the enforcement procedures adopted by the ADA mirror those of Title VII of the Civil Rights Act. A claimant under the ADA must file a claim with the EEOC within 180 days of the alleged discriminatory event, or within 300 days in states with approved enforcement agencies such as the Human Rights Commission. These are commonly called dual agency states or

Section 706 agencies. The EEOC has 180 days to investigate the allegation and sue the employer, or to issue a right-to-sue notice to the employee. The employee will have 90 days to file a civil action from the date of this notice.[31]

Safety professionals should be aware that the governing federal agency for the ADA is the Equal Employment Opportunity Commission. Enforcement of the ADA is also permitted by the attorney general or by private lawsuit. Remedies under these titles include the ordered modification of a facility, and civil penalties of up to $50,000 for the first violation and $100,000 for any subsequent violations. Section 505 permits reasonable attorney fees and litigation costs for the prevailing party in an ADA action, but under Section 513, Congress encourages the use of arbitration to resolve disputes arising under the ADA.[32]

With the passage of the Civil Rights Act of 1991, the remedies provided under the ADA were modified. Employment discrimination (whether intentional or by practice) that has a discriminatory effect on qualified individuals may include hiring, reinstatement, promotion, back pay, front pay, reasonable accommodation, or other actions that will make an individual "whole." Payment of attorney fees, expert witness fees, and court fees is still permitted, and jury trials are also allowed.

Compensatory and punitive damages are also made available if intentional discrimination is found. Damages may be available to compensate for actual monetary losses, future monetary losses, mental anguish, and inconvenience. Punitive damages are also available if an employer acted with malice or reckless indifference. The total amount of punitive and compensatory damages for future monetary loss and emotional injury for each individual is limited, and is based upon the size of the employer.

The ADA can be a very complex area for safety professionals given the potential entanglements within various safety functions. Safety professionals should acquire guidance from their human resources department or legal counsel before addressing any issue with ADA implications. As noted above, prudent safety professionals may consider simply asking "How can I help you?" and *listening* to the applicant or employee before acting in any manner. Careful assessment and evaluation can assist the safety professional in the successful navigation of the ADA waters.

CHAPTER QUESTIONS

1. To qualify for protection under the ADA, an individual must:
   a. Possess a permanent physical disability
   b. Possess a permanent mental disability
   c. Possess a record of the disability
   d. All of the above
2. The enforcement agency for the ADA is:
   a. Department of Labor
   b. Department of Defense
   c. Equal Employment Opportunity Commission
   d. None of the above
3. The "regard as having an impairment" prong of the test means:
   a. The individual does not have a permanent disability
   b. The individual is treated as if he or she had a disability

      c. The individual was discriminated against

      d. All of the above

  4. The ADA was amended by:

      a. The Civil Rights Act of 1964

      b. The Civil Rights Act of 1991

      c. The Bill of Rights

      d. None of the above

  5. A reasonable accommodation can include:

      a. A chair

      b. A schedule change

      c. A walkway

      d. All of the above

ANSWERS

  1. d

  2. c

  3. d

  4. b

  5. d

ENDNOTES

  1. National Employment Law Conference, 2010.
  2. ADA Section 305.
  3. ADA Section 102(a); 42 USC Section 12122.
  4. ADA Section 101(8).
  5. EEOC Interpretive Rules, 56 *Fed. Reg.* 35 (July 26, 1991).
  6. 42 *Fed. Reg.* 22686 (May 4, 1977); S. Rep. 101-116; H. Rep. 101-485, Part 2, 51.
  7. Subtitle A, Section 3(2).
  8. 28 CFR Section 41.31.
  9. See *Jasany v. U.S. Postal Service*, 755 F.2d 1244 (6th Cir. 1985).
  10. S. Rep. 101-116,23; H. Rep. 101-485, Part 2, 52-53.
  11. 45 CFR 84.3(j)(2)(iv).
  12. ADA, Title I, Section 101(8).
  13. EEOC Interpretive Rules, 56 *Fed. Reg.* 35 (July 26, 1991).
  14. Ibid.
  15. ADA, Section 103(b).
  16. EEOC Interpretive Guidelines, EEOC, 1994.
  17. Ibid.
  18. 56 *Fed. Reg.* 35745 (July 26, 1991).
  19. ADA Section 101(9).
  20. EEOC Interpretative Rules, EEOC, 1994.
  21. Ibid.
  22. S. Comm. on Lab. and Hum. Resources Rep. at 38; H. Comm. on Jud. Rep. at 42.
  23. Ibid.
  24. EEOC Interpretive Guidelines, 56 *Fed. Reg.* 35751 (July 26, 1991).
  25. ADA Section 102(c).
  26. ADA Section 511(b).

27. ADA Section 3(1).
28. H. Rep. 101-485, Part 2, 51.
29. Ibid.
30. EEOC Interpretive Guidelines, EEOC, 1994.
31. S. Rep. 101-116, 21; H. Rep. 101-485, Part 2, 51; Part 3, 28.
32. ADA Sections 505 and 513.

14 OSHA Inspections and Defenses

Learning objectives:

1. Acquire an understanding of the OSHA compliance inspection processes.
2. Acquire an understanding of the OSHA penalty schedule.
3. Acquire an understanding of the possible defenses to an OSHA citation.

The Occupational Safety and Health Act of 1970 (OSH Act) is the primary reason for the existence of most careers in the safety profession today. Established under the OSH Act, the Occupational Safety and Health Administration (OSHA) was developed as the enforcement arm of the OSH Act, and thus can and has penalized companies and organizations for safety violations through monetary fines or penalties. As most safety professionals are also aware, OSHA can also refer fatality and serious violations to the U.S. Department of Justice for criminal prosecution.

Safety professionals should be aware of the new Severe Violator Enforcement Program established in 2010, which "is intended to focus OSHA enforcement resources on recalcitrant employers who endanger workers by demonstrating indifference to their responsibilities under the law."[1] Additionally, safety professionals should be aware that it is proposed to increase the monetary penalties. "The current maximum penalty for a serious violation, one capable of causing death or serious physical harm, is only $7,000 and the maximum penalty for a willful violation is

$70,000. The average penalty for a serious violation will increase from about $1,000 to an average $3,000 to $4,000. Monetary penalties for violations of the OSH Act have been increased only once in 40 years despite inflation. The Protecting America's Workers Act would raise these penalties, for the first time since 1990, to $12,000 and $250,000, respectively. Future penalty increases would also be tied to inflation."[2]

Although currently in committee, safety professionals should be aware of the Protecting America's Workers Act of 2009[3] (PAWA), which proposes to substantially raise the civil and criminal penalties. As proposed, PAWA would increase the civil penalty for a willful or repeat OSHA violation to a minimum fine of $8,000 and a maximum fine of $120,000. Minimum and maximum fines for willful or repeat violations causing a fatality would increase to a $50,000 minimum and a $250,000 maximum. Serious violations would increase from $7,000 to $12,000, with a maximum of $50,000 if a fatality is involved. On the criminal side, the maximum penalty would be increased to a felony and a 10-year prison term.

PAWA also currently possesses a provision to adjust the monetary penalties every 4 years for inflation and expands the current scope of "employer" and enhances the current OSHA antiretaliation provisions. As of this writing, PAWA currently has fifty-one cosponsors in the House and Senate and is currently in the Subcommittee on Health, Education, Labor, and Pensions.

Although the day-to-day function of the safety professional does not involve any interaction with OSHA, once an OSHA compliance officer conducts a compliance inspection, the potential of receiving a citation for identified alleged violations increases and the safety professional's function will change considerably. Most citations issued by OSHA for alleged violations possess a categorization of the alleged violation, as well as a proposed monetary penalty for each alleged violation. Upon receipt of the citation, the safety professional is provided a very short time period (15 working days) to appeal it. Thus, it is important that safety professionals quickly identify the category of each alleged violation, as well as the proposed monetary penalty. The gravity of the violation is the primary factor in determining penalties.[4] In assessing the gravity of a violation, the compliance officer or area director must consider the severity of the injury or illness that could result, and the probability that an injury or illness could occur as a result of the violation.[5] Specific penalty assessment tables assist the area director or compliance officer in determining the appropriate fine for the violation.[6]

In general, the OSHA monetary penalty structure is classified according to the type and gravity of the particular violation. Violations of OSHA standards or the general duty clause are categorized as *de minimis*, other (nonserious), serious, repeat, and willful. Monetary penalties assessed by the secretary vary according to the degree of the violation. Penalties range from no monetary penalty to ten times the imposed penalty for repeat or willful violations.[7] Additionally, the secretary of labor may refer willful violations to the U.S. Department of Justice for imposition of criminal sanctions.[8]

Safety professionals should also be aware that the OSH Act provides criminal penalties in four circumstances.[9] In the first, anyone inside or outside of the Department of Labor or OSHA who gives advance notice of an inspection, without authority from the secretary, may be fined up to $1,000, imprisoned for up to 6 months, or both. Second, any employer or person who intentionally falsifies statements or OSHA records that must be prepared, maintained, or submitted under the OSH Act may, if found guilty, be fined up to $10,000, imprisoned for up to 6 months, or both. Third, any person responsible for a violation of an OSHA standard, rule, order, or regulation that causes the death of an employee may, upon conviction, be fined up to $10,000, imprisoned for up to 6 months, or both. If convicted for a second violation, punishment may be a fine of up to $20,000, imprisonment for up to 1 year, or both.[10] Finally, if an individual is convicted of forcibly resisting or assaulting a compliance officer or other Department of Labor personnel, a fine of $5,000, 3 years in prison, or both can be imposed. Any person convicted of killing a compliance officer or other OSHA or Department of Labor personnel, acting in his or her official capacity, may be sentenced to prison for any term of years or life.

Safety professionals must be aware that there is a 15-working-day limitation from the time the citation is received. If the time limitation is not met, there is a substantial probability that the company or organization will lose all appeal rights. Thus it is vital that safety professionals understand and fully explain their company or organization's rights to management in a timely fashion and file the notice of contest within 15 working days of the date the citation is received. Again, safety professionals must understand that failure to file the notice of contest in a timely manner can result in the loss of all appeal rights. Additionally, it is vitally important that safety professionals work with legal counsel before filing the notice of contest. Although the safety professional will be the company or organization's agent throughout the appeals process, and is often the "content-based expert" regarding the OSHA standards, defenses, and inspection, the appeal to ⌐ Occupational Safety and Health Review Commission (OSHRC) and courts is the domain of your legal counsel.

Safety professionals should be aware that once the citation is issued, OSHA has the burden of proof to prove each and every alleged violation. The preponderance of the evidence standard is utilized in OSHRC hearings, and Rules of Evidence are utilized throughout. In short, OSHA must prove each and every element of each and every alleged violation. In preparing the defenses, safety professionals should be looking for deficiencies or lack of proof for each and every alleged violation.

Working with legal counsel, the first level of possible defenses usually involves the procedural aspects of the citation. These defenses are usually very technical in nature, such as a defect in the inspection procedure, and are usually not successful. OSHA inspectors are usually well educated in the procedural requirements. Two general defenses in the procedural area have been utilized, including the statute of limitation defense and the lack of reasonable promptness defense. The statute of limitation defense is often utilized when the citation is issued beyond 6 months from the time of the alleged violation. In general, if the citation is not issued within 6 months of the date of the alleged violation, this is ground to dismiss the citation. The lack of reasonable promptness defense, although not an absolute defense, generally involves a similar delay in issuing the citation, which prejudices the employer in preparing the defense. Other procedural defenses may be available, depending on the circumstances.

Where the safety professional is invaluable to legal counsel in preparing the appeal is in the area of factual defenses, e.g., what happened before, during, and after the inspection. Factual defenses are usually based on what the circumstances were at the time of the alleged violation, what was or wasn't observed by the compliance officer, what actions or inactions involved the employees, and related facts. Factual defenses are based primarily on the circumstances. One of the most often utilized factually based defenses where the safety professional has a well-established safety and health program is the defense of unpreventable employee misconduct. To prove this defense, the safety professional must document each and every aspect of the inspection and assist legal counsel in providing evidence that the work rule or policy to prevent the violation from occurring is established; the work rule or policy was adequately communicated to all employees, including their supervisors or team leaders; the management team member took reasonable steps to discover the alleged violation; and the safety professional or other management team member, as the agent of the employer, effectively enforced the work rule or policy that the employees violated.

Safety professionals should provide legal counsel with written compliance programs, written safety policies, documentation of fair and consistent disciplinary action for violations, training and education documentation, as well as safety inspections, safety audits, and related documentation. Please be aware that some of the circuit courts require the defense of unpreventable employee misconduct to be plead as an affirmative defense prior to filing with the OSHRC.

Another common factual defense is the isolated incident defense. This defense is exactly what the title implies, namely, the employer possesses an adequate safety and health program and the alleged violation observed by the compliance officer was an isolated incident that is out of the norm of the safety and heath program and efforts of the employer. Safety professionals should be aware that in order to establish this defense, documentation and other supporting evidence must demonstrate the following:

1. The alleged violation resulted exclusively from the employee's conduct or misconduct.
2. The alleged violation was not "participated in, observed by or performed with the knowledge and/or consent of any supervisory personnel."
3. The employee's conduct or misconduct contravened a well-established company policy or work rule that was in effect at the time, well published to the employees, and actively enforced through disciplinary action or other appropriate procedures.[11]

It should be noted that the isolated incident defense is also an affirmative defense.

Depending on the facts involved in the alleged violation, safety professionals should be aware that the greater hazard in compliance defense may be applicable. In essence, this defense challenges the OSHA standard for the particular situation, arguing that compliance with the standard would create a greater hazard for employees than an alternative that is not in compliance with the promulgated standard. Safety and loss prevention professionals should be aware that this defense carries a substantial evidentiary burden to provide that the hazard of compliance is greater than the hazard of noncompliance, and that alternative means of protection were unavailable. Prudent safety professionals may want to explore the use of a variance in this type of situation before an inspection.

The defense of impossibility or infeasibility to comply is also available, depending on the circumstances. The basis of this defense is that achieving compliance with the standard is impossible because of the nature of the specific job or work. This defense is very narrow, and safety professionals should be aware that the burden shifts back to the employer to prove this defense, and there are two levels of proof, with the first being:

> It would have been technologically or economically infeasible to implement the standard's requirements under the circumstances, or compliance with the standard would have precluded performance of necessary work operations.[12]

Additionally, the safety professional must then prove that "the company used an alternative method of employee protection; or no feasible alternative means of protection was available."[13]

It should be noted that this defense carries with it a heavy burden of proof, and simply having difficulty achieving compliance or the fact that compliance may be inconvenient or expensive is usually not sufficient to sustain this burden. Safety professionals and legal counsel should be prepared to exhibit that all possible alternative forms of protection had been explored and tested, where feasible. Additionally, safety and loss prevention professionals should assemble documentation to prove that required work could not be properly performed if the standard had been utilized to achieve compliance. If the economically infeasible argument is to be made, safety professionals should be prepared to demonstrate that the cost is unreasonable in light of the protection afforded, and to show any adverse effects the cost would have on the business as a whole.[14]

A relatively simple defense that can be utilized, depending on the circumstances, is that of the machine was not in use. If the facts of the situation support that the machine cited by the compliance officer during the inspection was not in use and thus could not create the alleged violation identified in the citation, the defense that the machine or equipment was not in use may be viable.

The defense of the lack of the employer's knowledge can be argued in specific situations. Although this is often a difficult defense to prove because the employer should know the activities and operations within its facility, safety professionals may utilize this defense in unique situations to challenge OSHA's proof of the violation.

Safety professionals should be aware that the defenses identified above are not all of the available defenses. Creative safety professionals, working with legal counsel and their management team, can shape and create new defenses, depending on the facts and circumstances identified in the citation. Each and every detail in each and every alleged violation should be carefully scrutinized, with each possible defense reviewed and analyzed. Diligence, creativity, and an eye to detail will permit the safety professional and legal counsel to develop viable defenses to most alleged violations. If there are no defenses available, it may be time to attempt to negotiate reductions in the proposed categorization or monetary penalties through your good faith or other factors.

CHAPTER QUESTIONS

1. The maximum monetary penalty for a willful violation is _____.
2. The maximum penalty for a serious violation is _____.
3. PAWA is _____.
4. The maximum criminal sentence currently available under the OSH Act is _____.
5. Name at least one defense to an OSHA violation and explain.

ANSWERS

1. $70,000
2. $7,000
3. Protecting America's Workers Act
4. Six months
5. Isolated incident, machine not in operation, greater hazard, and other defenses.

ENDNOTES

1. OSHA News Release 10-538-NAT, April 22, 2010.
2. Ibid.
3. HR 2067, S. 1580.
4. OSHA Compliance Field Manual at XI-c3c (April 1977).
5. Ibid.
6. Ibid.
7. 29 USC Section 658(b).
8. Ibid. at (e).
9. 29 USC Section 666(e)–(g). See also, OSHA Manual, supra note 1 at VI-B.
10. A repeat criminal conviction doubled the possible criminal penalties.
11. Ibid.
12. Ibid.
13. Ibid.
14. Ibid.

15 New Cases, Laws, and Legislation

The minute you read something you can't understand, you can almost be sure it was drawn up by a lawyer.

—**Will Rogers**

Lawsuit: A machine which you go into as a pig and come out of as a sausage.

—**Ambrose Bierce**

Learning objectives:

1. To acquire an understanding of the new laws and regulations recently enacted.
2. To acquire an understanding of the proposed legislation potentially impacting the safety function.

Safety professionals should keep a careful eye on the proposed legislation and recent court decisions that potentially could impact the safety function, their industry, and the safety profession. In the past few years, safety professionals have experienced the historic election of President Obama in 2008, the Democrats possessing a filibuster-proof majority in the U.S. Senate, and one of the most substantial recessions/depressions experienced by the United States in the past 50 years. As most safety professionals are aware, a number of bills were proposed, including the Employee Free Choice Act, which possessed the ability to substantially change the labor and employment as well as safety landscape in the average American workplace.

Safety professionals should be aware that in 2009, the Obama administration substantially increased funding for many federal agencies, including OSHA. Secretary of Labor Hilda Solis realigned the Department of Labor's priorities, shifting resources to OSHA and other agencies charged with enforcement of workplace safety and health, as well as the enforcement of the Fair Labor Standards Act (FLSA).[1] Through this shift in resources, OSHA was able to add 177 new compliance officer positions with the anticipation of additional positions in the future. Additionally, OSHA's budget was increased by $46.5 million, providing a budget of $554.6 million, with a proposed increase for 2011.

Safety professionals should also be aware that in the 2010 consolidated appropriations bill,[2] which President Obama signed on December 16, 2009, the Department of Labor's budget was increased to $13.3 billion, and the Wage and Hour Division of the Department of Labor received an increase of $35 million. Given these budget increases, safety professionals should know that the Department of Labor added nearly 700 new enforcement staff personnel, tasked with enforcement of OSHA, wage and hour, and other employment-related laws. For 2011, the Obama administration is proposing an additional $14 billion for the Department of Labor, $20 million for the Wage and Hour Division, and $14 million for OSHA. With these increases in budget amounts and enforcement personnel, safety professionals should be prepared for increased and aggressive enforcement of OSHA standards in the workplace.

Although there were a number of standards promulgated by OSHA in recent years, the most potentially impacting change has been in the Severe Violator Enforcement Program (SVEP). This directive concentrated OSHA's resources on inspecting companies or organizations that have "demonstrated indifference to their OSHA obligations by willful, repeated or failure-to-abate violations."[3] This directive also increased the potential monetary penalties by 10% for high-gravity serious, willful, repeat, or failure-to-abate violations received by the company or organization within the last 5 years.

Safety professionals should also be aware that OSHA has announced an airborne infectious disease rule that may impact a substantial number of workplaces. This proposed rule is modeled after the Cal-OSHA aerosol transmissible disease rule, which required respiratory protection, fit testing, disease exposure control plans, medical surveillance, and communication procedures, among other requirements.

Safety professionals should be aware that the Protecting American Workers Act of 2009 (PAWA) is currently in the Senate Health, Education, Labor, and Pension (HELP) committee and possesses a substantial probability of being reintroduced in the next congressional session. Safety professionals should know that the original PAWA bill was cosponsored by then Senator Obama (now President Obama), then Senator Biden (now Vice President Biden), and then Senator Clinton (now Secretary of State Clinton). As discussed in greater detail in Chapter 14, the proposed PAWA legislation would extend OSHA's jurisdiction to all public and private sector employees, expand the scope of the covered employer, enhance the antiretaliation provisions of the OSH Act, and substantially increase the monetary and criminal penalties.

The proposed legislation acquiring the most publicity and which will definitely impact the safety function, especially in nonunion operations, is the proposed Employee Free Choice Act (EFCA). Although this proposed legislation has changed from the original version, wherein the National Labor Relations Act (NLRA) would be amended to require the National Labor Relations Board (NLRB) to certify a union as the representative if 51% of the employees signed authorization cards, the current version of this proposed law has dropped the card check provisions but added a 5- to 10-day requirement for an election after the union acquired 30% of the authorization cards in favor of the union. Safety professionals should also be aware that the current proposed version of EFCA increases the labor law penalties for labor violations and possesses a required arbitration provision to settle contract negotiation disputes.

EFCA is controversial and safety professionals should keep an eye on this legislation, especially changes or modifications to the NLRA that the EFCA may require. If passed, safety professionals should be aware that the EFCA should create an easier path for labor organizations to campaign and organize nonunion companies and organizations. IF EFCA becomes law, safety professionals working in nonunion companies can anticipate substantially more organizing campaigns by labor organizations.

In the wage and hour arena, safety professionals should be aware of the proposed Paycheck Fairness Act of 2009 (PFA), which amended the Equal Pay Act by addressing the Equal Pay Act's "other factors" exceptions, which provided a general prohibition on wage differentials between men and women in the same company performing the same or similar work.[4] Safety professionals should also be aware of the proposed Family-Friendly Workplace Act, which would amend the FLSA to permit companies or organizations to offer "comp time" off at the overtime rate of 1½ hours.[5] And safety professionals should be aware of the Breastfeeding Promotion Act[6] and the Working for Adequate Gains in Employment in Service Act, addressing minimum wage for employees receiving tips as part of their wages.[7]

Within the FMLA arena, safety professionals should be aware of the proposed Balancing Act of 2009,[8] which would amend and increase the scope of the FMLA to include 12 weeks of paid leave for family issues, 7 days of paid sick leave, parental involvement leave, domestic violence leave, and other expansions of the current FMLA. Under the proposed Healthy Family Act,[9] companies and organizations would be required to provide 1 week of paid sick or health-related leave each year.

Safety professionals should be aware that there are a number of proposed amendments to the FMLA, including the Family and Medical Leave Enhancement Act of 2009,[10] Paid Vacation Act,[11] Domestic Violence Leave Act,[12] Military Family Leave Act of 2009,[13] and Living Organ Donor Job Security Act,[14] among a number of proposed legislative changes that can impact the safety function if passed.

Within the area of antidiscrimination legislation, safety professionals should be aware of a number of proposed amendments to Title VII and related laws, and proposed new laws, including the Employment Nondiscrimination Act of 2009,[15] which prohibits discrimination against gay, bisexual, and transgender employees, and the Protecting Older Workers against Discrimination Act,[16] which proposes to address recent U.S. Supreme Court decisions regarding "but for" causation for age. Safety professionals should be aware that there is also proposed legislation that would prohibit companies or organizations from basing adverse employment decisions on the employee's credit report[17] and addressing worker classification as subcontractors.[18]

As can be seen, safety professionals should diligently keep track of the proposed legislation that could directly or indirectly impact their safety programs and activities. With the increase in budgetary amounts and addition of a large number of compliance-related positions within OSHA and other federal governmental agencies, safety professionals can expect a higher volume of compliance-related inspections, and thus the probability of an increase of alleged violations and monetary penalties. Prudent safety professionals should initiate or continue their proactive systematic approach to address all compliance-related issues far in advance of any compliance-related inspection activities and maintain a level of knowledge with regards to any and all new laws that could impact their safety function.

CHAPTER QUESTIONS (TRUE/FALSE)

1. OSHA's funding decreased in 2009.
2. EFCA stands for Employee Freedom Classification Act.
3. There are no proposed laws that would impact FMLA.
4. The Paycheck Fairness Act addresses age discrimination.
5. PAWA stands for Protecting American Workers Act.

ANSWERS

1. False
2. False
3. False
4. False
5. True

ENDNOTES

1. Department of Labor website, Media Report Annual 2009, at www.dol.gov.
2. HR 3288.
3. Memorandum—Administrative Enhancement to OSHA's Penalty Policy, April 22, 2010.
4. HR 12, S. 182 (2010).
5. HR 933 (2010).
6. HR 2819 (2010).
7. HR 2570 (2010).
8. HR 3047 (2009).
9. SB 1152 (2009).
10. HR 824 (2009).
11. HR 2564 (2009).
12. HR 2515 (2009).
13. S. 1441 (2009).
14. HR 2772 (2009).
15. HR 3017, S. 1373 (2009).
16. S. 1756, HR 3721 (2009).
17. Equal Employment for All Act, HR 3149 (2009).
18. Taxpayer Responsibility, Accountability, and Consistency Act of 2009, S. 2882 (2009).

Appendix 1: National Labor Relations Act

Congress enacted the National Labor Relations Act (NLRA) in 1935 to protect the rights of employees and employers, to encourage collective bargaining, and to curtail certain private sector labor and management practices, which can harm the general welfare of workers, businesses, and the U.S. economy.

NATIONAL LABOR RELATIONS ACT

Also cited NLRA or the Act; 29 U.S.C. §§ 151–169
[Title 29, Chapter 7, Subchapter II, United States Code]

FINDINGS AND POLICIES

Section 1. [§151.] The denial by some employers of the right of employees to organize and the refusal by some employers to accept the procedure of collective bargaining lead to strikes and other forms of industrial strife or unrest, which have the intent or the necessary effect of burdening or obstructing commerce by (a) impairing the efficiency, safety, or operation of the instrumentalities of commerce; (b) occurring in the current of commerce; (c) materially affecting, restraining, or controlling the flow of raw materials or manufactured or processed goods from or into the channels of commerce, or the prices of such materials or goods in commerce; or (d) causing diminution of employment and wages in such volume as substantially to impair or disrupt the market for goods flowing from or into the channels of commerce.

The inequality of bargaining power between employees who do not possess full freedom of association or actual liberty of contract and employers who are organized in the corporate or other forms of ownership association substantially burdens and affects the flow of commerce, and tends to aggravate recurrent business depressions, by depressing wage rates and the purchasing power of wage earners in industry and by preventing the stabilization of competitive wage rates and working conditions within and between industries.

Experience has proved that protection by law of the right of employees to organize and bargain collectively safeguards commerce from injury, impairment, or interruption, and promotes the flow of commerce by removing certain recognized sources of industrial strife and unrest, by encouraging practices fundamental to the friendly adjustment of industrial disputes arising out of differences as to wages, hours, or other working conditions, and by restoring equality of bargaining power between employers and employees.

Experience has further demonstrated that certain practices by some labor organizations, their officers, and members have the intent or the necessary effect of burdening or obstructing commerce by preventing the free flow of goods in such commerce through strikes and other forms of industrial unrest or through concerted activities which impair the interest of the public in the free flow of such commerce. The elimination of such practices is a necessary condition to the assurance of the rights herein guaranteed.

It is declared to be the policy of the United States to eliminate the causes of certain substantial obstructions to the free flow of commerce and to mitigate and eliminate these obstructions when they have occurred by encouraging the practice and procedure of collective bargaining and by protecting the exercise by workers of full freedom of association, self-organization, and designation of representatives of their own choosing, for the purpose of negotiating the terms and conditions of their employment or other mutual aid or protection.

DEFINITIONS

Sec. 2. [§152.] When used in this Act [subchapter]—

(1) The term "person" includes one or more individuals, labor organizations, partnerships, associations, corporations, legal representatives, trustees, trustees in cases under title 11 of the United States Code [under title 11], or receivers.

(2) The term "employer" includes any person acting as an agent of an employer, directly or indirectly, but shall not include the United States or any wholly owned Government corporation, or any Federal Reserve Bank, or any State or political subdivision thereof, or any person subject to the Railway Labor Act [45 U.S.C. § 151 et seq.], as amended from time to time, or any labor organization (other than when acting as an employer), or anyone acting in the capacity of officer or agent of such labor organization.

[Pub. L. 93-360, § 1(a), July 26, 1974, 88 Stat. 395, deleted the phrase "or any corporation or association operating a hospital, if no part of the net earnings inures to the benefit of any private shareholder or individual" from the definition of "employer."]

(3) The term "employee" shall include any employee, and shall not be limited to the employees of a particular employer, unless the Act [this subchapter] explicitly states otherwise, and shall include any individual whose work has ceased as a consequence of, or in connection with, any current labor dispute or because of any unfair labor practice, and who has not obtained any other regular and substantially equivalent employment, but shall not include any individual employed as an agricultural laborer, or in the domestic service of any family or person at his home, or any individual employed by his parent or spouse, or any

individual having the status of an independent contractor, or any individual employed as a supervisor, or any individual employed by an employer subject to the Railway Labor Act [45 U.S.C. § 151 et seq.], as amended from time to time, or by any other person who is not an employer as herein defined.

(4) The term "representatives" includes any individual or labor organization.

(5) The term "labor organization" means any organization of any kind, or any agency or employee representation committee or plan, in which employees participate and which exists for the purpose, in whole or in part, of dealing with employers concerning grievances, labor disputes, wages, rates of pay, hours of employment, or conditions of work.

(6) The term "commerce" means trade, traffic, commerce, transportation, or communication among the several States, or between the District of Columbia or any Territory of the United States and any State or other Territory, or between any foreign country and any State, Territory, or the District of Columbia, or within the District of Columbia or any Territory, or between points in the same State but through any other State or any Territory or the District of Columbia or any foreign country.

(7) The term "affecting commerce" means in commerce, or burdening or obstructing commerce or the free flow of commerce, or having led or tending to lead to a labor dispute burdening or obstructing commerce or the free flow of commerce.

(8) The term "unfair labor practice" means any unfair labor practice listed in section 8 [section 158 of this title].

(9) The term "labor dispute" includes any controversy concerning terms, tenure or conditions of employment, or concerning the association or representation of persons in negotiating, fixing, maintaining, changing, or seeking to arrange terms or conditions of employment, regardless of whether the disputants stand in the proximate relation of employer and employee.

(10) The term "National Labor Relations Board" means the National Labor Relations Board provided for in section 3 of this Act [section 153 of this title].

(11) The term "supervisor" means any individual having authority, in the interest of the employer, to hire, transfer, suspend, lay off, recall, promote, discharge, assign, reward, or discipline other employees, or responsibly to direct them, or to adjust their grievances, or effectively to recommend such action, if in connection with the foregoing the exercise of such authority is not of a merely routine or clerical nature, but requires the use of independent judgment.

(12) The term "professional employee" means—

    (a) any employee engaged in work (i) predominantly intellectual and varied in character as opposed to routine mental, manual, mechanical, or physical work; (ii) involving the consistent exercise of discretion and judgment in its performance; (iii) of such a character that the output produced or the result accomplished cannot be standardized in relation to a given period of time; (iv) requiring knowledge of an advanced type in a field of science or learning customarily acquired by a prolonged course of specialized intellectual instruction and study in an institution of higher learning or a hospital, as distinguished from a general academic education or from an apprenticeship or from training in the performance of routine mental, manual, or physical processes; or

(b)   any employee, who (i) has completed the courses of specialized intellectual instruction and study described in clause (iv) of paragraph (a), and (ii) is performing related work under the supervision of a professional person to qualify himself to become a professional employee as defined in paragraph (a).

(13)   In determining whether any person is acting as an "agent" of another person so as to make such other person responsible for his acts, the question of whether the specific acts performed were actually authorized or subsequently ratified shall not be controlling.

(14)   The term "health care institution" shall include any hospital, convalescent hospital, health maintenance organization, health clinic, nursing home, extended care facility, or other institution devoted to the care of sick, infirm, or aged person.

[Pub. L. 93-360, § 1(b), July 26, 1974, 88 Stat. 395, added par. (14).]

NATIONAL LABOR RELATIONS BOARD

Sec. 3. [§ 153.]

(a)   **[Creation, composition, appointment, and tenure; Chairman; removal of members]** The National Labor Relations Board (hereinafter called the "Board") created by this Act [subchapter] prior to its amendment by the Labor Management Relations Act, 1947 [29 U.S.C. § 141 et seq.], is continued as an agency of the United States, except that the Board shall consist of five instead of three members, appointed by the President by and with the advice and consent of the Senate. Of the two additional members so provided for, one shall be appointed for a term of five years and the other for a term of two years. Their successors, and the successors of the other members, shall be appointed for terms of five years each, excepting that any individual chosen to fill a vacancy shall be appointed only for the unexpired term of the member whom he shall succeed. The President shall designate one member to serve as Chairman of the Board. Any member of the Board may be removed by the President, upon notice and hearing, for neglect of duty or malfeasance in office, but for no other cause.

(b)   **[Delegation of powers to members and regional directors; review and stay of actions of regional directors; quorum; seal]** The Board is authorized to delegate to any group of three or more members any or all of the powers which it may itself exercise. The Board is also authorized to delegate to its regional directors its powers under section 9 [section 159 of this title] to determine the unit appropriate for the purpose of collective bargaining, to investigate and provide for hearings, and determine whether a question of representation exists, and to direct an election or take a secret ballot under subsection (c) or (e) of section 9 [section 159 of this title] and certify the results thereof, except that upon the filling of a request therefore with the Board by any interested person, the Board may review any action of a regional director delegated to him under this paragraph, but such a review shall not, unless specifically ordered by the Board, operate as a stay of any action taken by the regional director. A vacancy in the Board shall not impair the right of the remaining members to exercise all of the powers of the Board, and three members of the Board shall, at all times, constitute a quorum of the Board, except that two members shall constitute a quorum of any group designated pursuant to the first sentence hereof. The Board shall have an official seal which shall be judicially noticed.

(c) **[Annual reports to Congress and the President]** The Board shall at the close of each fiscal year make a report in writing to Congress and to the President summarizing significant case activities and operations for that fiscal year.

(d) **[General Counsel; appointment and tenure; powers and duties; vacancy]** There shall be a General Counsel of the Board who shall be appointed by the President, by and with the advice and consent of the Senate, for a term of four years. The General Counsel of the Board shall exercise general supervision over all attorneys employed by the Board (other than administrative law judges and legal assistants to Board members) and over the officers and employees in the regional offices. He shall have final authority, on behalf of the Board, in respect of the investigation of charges and issuance of complaints under section 10 [section 160 of this title], and in respect of the prosecution of such complaints before the Board, and shall have such other duties as the Board may prescribe or as may be provided by law. In case of vacancy in the office of the General Counsel the President is authorized to designate the officer or employee who shall act as General Counsel during such vacancy, but no person or persons so designated shall so act (1) for more than forty days when the Congress is in session unless a nomination to fill such vacancy shall have been submitted to the Senate, or (2) after the adjournment sine die of the session of the Senate in which such nomination was submitted.

[The title "administrative law judge" was adopted in 5 U.S.C. § 3105.]

Sec. 4. [§ 154. Eligibility for reappointment; officers and employees; payment of expenses]

(a) Each member of the Board and the General Counsel of the Board shall be eligible for reappointment, and shall not engage in any other business, vocation, or employment. The Board shall appoint an executive secretary, and such attorneys, examiners, and regional directors, and such other employees as it may from time to time find necessary for the proper performance of its duties. The Board may not employ any attorneys for the purpose of reviewing transcripts of hearings or preparing drafts of opinions except that any attorney employed for assignment as a legal assistant to any Board member may for such Board member review such transcripts and prepare such drafts. No administrative law judge's report shall be reviewed, either before or after its publication, by any person other than a member of the Board or his legal assistant, and no administrative law judge shall advise or consult with the Board with respect to exceptions taken to his findings, rulings, or recommendations. The Board may establish or utilize such regional, local, or other agencies, and utilize such voluntary and uncompensated services, as may from time to time be needed. Attorneys appointed under this section may, at the direction of the Board, appear for and represent the Board in any case in court. Nothing in this Act [subchapter] shall be construed to authorize the Board to appoint individuals for the purpose of conciliation or mediation, or for economic analysis.

[The title "administrative law judge" was adopted in 5 U.S.C. § 3105.]

(b) All of the expenses of the Board, including all necessary traveling and subsistence expenses outside the District of Columbia incurred by the members or employees of the Board under its orders, shall be allowed and paid on the presentation of itemized vouchers therefore approved by the Board or by any individual it designates for that purpose.

Sec. 5. [§ 155. Principal office, conducting inquiries throughout country; participation in decisions or inquiries conducted by member] The principal office of the Board shall be in the District of Columbia, but it may meet and exercise any or all of its powers at any other place. The Board may, by one or more of its members or by such agents or agencies as it may designate, prosecute any inquiry necessary to its functions in any part of the United States. A member who participates in such an inquiry shall not be disqualified from subsequently participating in a decision of the Board in the same case.

Sec. 6. [§ 156. Rules and regulations] The Board shall have authority from time to time to make, amend, and rescind, in the manner prescribed by the Administrative Procedure Act [by subchapter II of chapter 5 of title 5], such rules and regulations as may be necessary to carry out the provisions of this Act [subchapter].

RIGHTS OF EMPLOYEES

Sec. 7. [§ 157.] Employees shall have the right to self-organization, to form, join, or assist labor organizations, to bargain collectively through representatives of their own choosing, and to engage in other concerted activities for the purpose of collective bargaining or other mutual aid or protection, and shall also have the right to refrain from any or all such activities except to the extent that such right may be affected by an agreement requiring membership in a labor organization as a condition of employment as authorized in section 8(a)(3) [section 158(a)(3) of this title].

Unfair Labor Practices

Sec. 8. [§ 158.] (a) [Unfair labor practices by employer] It shall be an unfair labor practice for an employer—

(1) to interfere with, restrain, or coerce employees in the exercise of the rights guaranteed in section 7 [section 157 of this title];

(2) to dominate or interfere with the formation or administration of any labor organization or contribute financial or other support to it: *Provided*, That subject to rules and regulations made and published by the Board pursuant to section 6 [section 156 of this title], an employer shall not be prohibited from permitting employees to confer with him during working hours without loss of time or pay;

(3) by discrimination in regard to hire or tenure of employment or any term or condition of employment to encourage or discourage membership in any labor organization: *Provided*, That nothing in this Act [subchapter], or in any other statute of the United States, shall preclude an employer from making an agreement with a labor organization (not established, maintained, or assisted by any action defined in section 8(a) of this Act [in this subsection] as an unfair labor practice) to require as a condition of employment membership therein on or after the thirtieth day following the beginning of such employment or the effective date of such agreement, whichever is the later, (i) if such labor organization is the representative of the employees as provided in section 9(a) [section 159(a) of this title], in the appropriate collective-bargaining unit covered by such agreement when made, and (ii) unless following an election held as provided in section 9(e) [section 159(e) of this title] within one year preceding the effective date of such agreement, the Board shall have certified that at least a majority of the employees eligible to vote in such election have voted to rescind the authority of such labor organization to make such an agreement: *Provided further*, That no employer shall justify any discrimination against an employee for non-membership in a labor organization (A) if he has reasonable grounds for believing that such membership was not available to the employee on the same terms and conditions generally applicable to other members, or (B) if he has reasonable grounds for believing that membership was denied or terminated for reasons other than the failure of the employee to tender the periodic dues and the initiation fees uniformly required as a condition of acquiring or retaining membership;

(4) to discharge or otherwise discriminate against an employee because he has filed charges or given testimony under this Act [subchapter];

(5) to refuse to bargain collectively with the representatives of his employees, subject to the provisions of section 9(a) [section 159(a) of this title].

(b) **[Unfair labor practices by labor organization]** It shall be an unfair labor practice for a labor organization or its agents—

(1) to restrain or coerce (A) employees in the exercise of the rights guaranteed in section 7 [section 157 of this title]: *Provided*, That this paragraph shall not impair the right of a labor organization to prescribe its own rules with respect to the acquisition or retention of membership therein; or (B) an employer in the selection of his representatives for the purposes of collective bargaining or the adjustment of grievances;

(2) to cause or attempt to cause an employer to discriminate against an employee in violation of subsection (a)(3) [of subsection (a)(3) of this section] or to discriminate against an employee with respect to whom membership in such organization has been denied or terminated on some ground other than his failure to tender the periodic dues and the initiation fees uniformly required as a condition of acquiring or retaining membership;

(3) to refuse to bargain collectively with an employer, provided it is the representative of his employees subject to the provisions of section 9(a) [section 159(a) of this title];

(4) (i) to engage in, or to induce or encourage any individual employed by any person engaged in commerce or in an industry affecting commerce to engage in, a strike or a refusal in the course of his employment to use, manufacture, process, transport, or otherwise handle or work on any goods, articles, materials, or commodities or to perform any services; or (ii) to threaten, coerce, or restrain any person engaged in commerce or in an industry affecting commerce, where in either case an object thereof is—

(A) forcing or requiring any employer or self-employed person to join any labor or employer organization or to enter into any agreement which is prohibited by section 8(e) [subsection (e) of this section];

(B) forcing or requiring any person to cease using, selling, handling, transporting, or otherwise dealing in the products of any other producer, processor, or manufacturer, or to cease doing business with any other person, or forcing or requiring any other employer to recognize or bargain with a labor organization as the representative of his employees unless such labor organization has been certified as the representative of such employees under the provisions of section 9 [section 159 of this title]: *Provided*, That nothing contained in this clause (B) shall be construed to make unlawful, where not otherwise unlawful, any primary strike or primary picketing;

(C) forcing or requiring any employer to recognize or bargain with a particular labor organization as the representative of his employees if another labor organization has been certified as the representative of such employees under the provisions of section 9 [section 159 of this title];

   (D)   forcing or requiring any employer to assign particular work to employees in a particular labor organization or in a particular trade, craft, or class rather than to employees in another labor organization or in another trade, craft, or class, unless such employer is failing to conform to an order or certification of the Board determining the bargaining representative for employees performing such work:

      Provided, That nothing contained in this subsection (b) [this subsection] shall be construed to make unlawful a refusal by any person to enter upon the premises of any employer (other than his own employer), if the employees of such employer are engaged in a strike ratified or approved by a representative of such employees whom such employer is required to recognize under this Act [subchapter]: *Provided further*, That for the purposes of this paragraph (4) only, nothing contained in such paragraph shall be construed to prohibit publicity, other than picketing, for the purpose of truthfully advising the public, including consumers and members of a labor organization, that a product or products are produced by an employer with whom the labor organization has a primary dispute and are distributed by another employer, as long as such publicity does not have an effect of inducing any individual employed by any person other than the primary employer in the course of his employment to refuse to pick up, deliver, or transport any goods, or not to perform any services, at the establishment of the employer engaged in such distribution;

  (5) to require of employees covered by an agreement authorized under subsection (a)(3) [of this section] the payment, as a condition precedent to becoming a member of such organization, of a fee in an amount which the Board finds excessive or discriminatory under all the circumstances. In making such a finding, the Board shall consider, among other relevant factors, the practices and customs of labor organizations in the particular industry, and the wages currently paid to the employees affected;

  (6) to cause or attempt to cause an employer to pay or deliver or agree to pay or deliver any money or other thing of value, in the nature of an exaction, for services which are not performed or not to be performed; and

  (7) to picket or cause to be picketed, or threaten to picket or cause to be picketed, any employer where an object thereof is forcing or requiring an employer to recognize or bargain with a labor organization as the representative of his employees, or forcing or requiring the

employees of an employer to accept or select such labor organization as their collective-bargaining representative, unless such labor organization is currently certified as the representative of such employees:

(A) where the employer has lawfully recognized in accordance with this Act [subchapter] any other labor organization and a question concerning representation may not appropriately be raised under section 9(c) of this Act [section 159(c) of this title],

(B) where within the preceding twelve months a valid election under section 9(c) of this Act [section 159(c) of this title] has been conducted, or

(C) where such picketing has been conducted without a petition under section 9(c) [section 159(c) of this title] being filed within a reasonable period of time not to exceed thirty days from the commencement of such picketing: *Provided*, That when such a petition has been filed the Board shall forthwith, without regard to the provisions of section 9(c)(1) [section 159(c)(1) of this title] or the absence of a showing of a substantial interest on the part of the labor organization, direct an election in such unit as the Board finds to be appropriate and shall certify the results thereof: *Provided further*, That nothing in this subparagraph (C) shall be construed to prohibit any picketing or other publicity for the purpose of truthfully advising the public (including consumers) that an employer does not employ members of, or have a contract with, a labor organization, unless an effect of such picketing is to induce any individual employed by any other person in the course of his employment, not to pick up, deliver or transport any goods or not to perform any services.

Nothing in this paragraph (7) shall be construed to permit any act which would otherwise be an unfair labor practice under this section 8(b) [this subsection].

(c) **[Expression of views without threat of reprisal or force or promise of benefit]** The expressing of any views, argument, or opinion, or the dissemination thereof, whether in written, printed, graphic, or visual form, shall not constitute or be evidence of an unfair labor practice under any of the provisions of this Act [subchapter], if such expression contains no threat of reprisal or force or promise of benefit.

(d) **[Obligation to bargain collectively]** For the purposes of this section, to bargain collectively is the performance of the mutual obligation of the employer and the representative of the employees to meet at reasonable times and confer in good faith with respect to wages, hours, and other terms and conditions of employment, or the negotiation of an agreement or any question arising thereunder, and the execution of a written contract incorporating any agreement reached if requested by either party, but such obligation does not compel either party to agree to a proposal or require the making of a concession: *Provided*, That where there is in effect a collective-bargaining contract covering employees in an industry affecting commerce, the duty to bargain collectively shall also mean that no party to such contract shall terminate or modify such contract, unless the party desiring such termination or modification—

(1) serves a written notice upon the other party to the contract of the proposed termination or modification sixty days prior to the expiration date thereof, or in the event such contract contains no expiration date, sixty days prior to the time it is proposed to make such termination or modification;

(2) offers to meet and confer with the other party for the purpose of negotiating a new contract or a contract containing the proposed modifications;

(3) notifies the Federal Mediation and Conciliation Service within thirty days after such notice of the existence of a dispute, and simultaneously therewith notifies any State or Territorial agency established to mediate and conciliate disputes within the State or Territory where the dispute occurred, provided no agreement has been reached by that time; and

(4) continues in full force and effect, without resorting to strike or lockout, all the terms and conditions of the existing contract for a period of sixty days after such notice is given or until the expiration date of such contract, whichever occurs later:

The duties imposed upon employers, employees, and labor organizations by paragraphs (2), (3), and (4) [paragraphs (2) to (4) of this subsection] shall become inapplicable upon an intervening certification of the Board, under which the labor organization or individual, which is a party to the contract, has been superseded as or ceased to be the representative of the employees subject to the provisions of section 9(a) [section 159(a) of this title], and the duties so imposed shall not be construed as requiring either party to discuss or agree to any modification of the terms and conditions contained in a contract for a fixed period, if such modification is to become effective before such terms and conditions can be reopened under the provisions of the contract. Any employee who engages in a strike within any notice period specified in this subsection, or who engages in any strike within the appropriate period specified in subsection (g) of this section, shall lose his status as an employee of the employer engaged in the particular labor dispute, for the purposes of sections 8, 9, and 10 of this Act [sections 158, 159, and 160 of this title], but such loss of status for such employee shall terminate if and when he is re-employed by such employer. Whenever the collective bargaining involves employees of a health care institution, the provisions of this section 8(d) [this subsection] shall be modified as follows:

(A)   The notice of section 8(d)(1) [paragraph (1) of this subsection] shall be ninety days; the notice of section 8(d)(3) [paragraph (3) of this subsection] shall be sixty days; and the contract period of section 8(d)(4) [paragraph (4) of this subsection] shall be ninety days.

(B)   Where the bargaining is for an initial agreement following certification or recognition, at least thirty days' notice of the existence of a dispute shall be given by the labor organization to the agencies set forth in section 8(d)(3) [in paragraph (3) of this subsection].

(C)   After notice is given to the Federal Mediation and Conciliation Service under either clause (A) or (B) of this sentence, the Service shall promptly communicate with the parties and use its best efforts, by mediation and conciliation, to bring them to agreement. The parties shall participate fully and promptly in such meetings as may be undertaken by the Service for the purpose of aiding in a settlement of the dispute.

[Pub. L. 93-360, July 26, 1974, 88 Stat. 395, amended the last sentence of Sec. 8(d) by striking the words "the sixty-day" and inserting the words "any notice" and by inserting before the words "shall lose" the phrase ", or who engages in any strike within the appropriate period specified in subsection (g) of this section." It also amended the end of paragraph Sec. 8(d) by adding a new sentence "Whenever the collective bargaining ... aiding in a settlement of the dispute."]

(e)   **[Enforceability of contract or agreement to boycott any other employer; exception]** It shall be an unfair labor practice for any labor organization and any employer to enter into any contract or agreement, express or implied, whereby such employer ceases or refrains or agrees to cease or refrain from handling, using, selling, transporting or otherwise dealing in any of the products of any other employer, or cease doing business with any other person, and any contract or agreement entered into heretofore or hereafter containing such an agreement shall be to such extent unenforceable and void: *Provided*, That nothing in this subsection (e) [this subsection] shall apply to an agreement between a labor organization and an employer in the construction industry relating to the contracting or subcontracting of work to be done at the site of the construction, alteration, painting, or repair of a building, structure, or other work: *Provided further*, That for the purposes of this subsection (e) and section 8(b)(4)(B) [this subsection and subsection (b)(4)(B) of this section] the terms "any employer," "any person engaged in commerce or an industry affecting commerce," and "any person" when used in relation to the terms "any other producer, processor, or manufacturer," "any other employer," or "any other person" shall not include persons in the relation of a jobber, manufacturer, contractor, or subcontractor working on the goods or premises of the jobber or manufacturer or performing parts of an integrated process of production in the apparel and clothing industry: *Provided further,* That nothing in this Act [subchapter] shall prohibit the enforcement of any agreement which is within the foregoing exception.

(f) **[Agreements covering employees in the building and construction industry]** It shall not be an unfair labor practice under subsections (a) and (b) of this section for an employer engaged primarily in the building and construction industry to make an agreement covering employees engaged (or who, upon their employment, will be engaged) in the building and construction industry with a labor organization of which building and construction employees are members (not established, maintained, or assisted by any action defined in section 8(a) of this Act [subsection (a) of this section] as an unfair labor practice) because (1) the majority status of such labor organization has not been established under the provisions of section 9 of this Act [section 159 of this title] prior to the making of such agreement, or (2) such agreement requires as a condition of employment, membership in such labor organization after the seventh day following the beginning of such employment or the effective date of the agreement, whichever is later, or (3) such agreement requires the employer to notify such labor organization of opportunities for employment with such employer, or gives such labor organization an opportunity to refer qualified applicants for such employment, or (4) such agreement specifies minimum training or experience qualifications for employment or provides for priority in opportunities for employment based upon length of service with such employer, in the industry or in the particular geographical area: *Provided,* That nothing in this subsection shall set aside the final proviso to section 8(a)(3) of this Act [subsection (a)(3) of this section]: *Provided further,* That any agreement which would be invalid, but for clause (1) of this subsection, shall not be a bar to a petition filed pursuant to section 9(c) or 9(e) [section 159(c) or 159(e) of this title].

(g) **[Notification of intention to strike or picket at any health care institution]** A labor organization before engaging in any strike, picketing, or other concerted refusal to work at any health care institution shall, not less than ten days prior to such action, notify the institution in writing and the Federal Mediation and Conciliation Service of that intention, except that in the case of bargaining for an initial agreement following certification or recognition the notice required by this subsection shall not be given until the expiration of the period specified in clause (B) of the last sentence of section 8(d) of this Act [subsection (d) of this section]. The notice shall state the date and time that such action will commence. The notice, once given, may be extended by the written agreement of both parties.

[Pub. L. 93-360, July 26, 1974, 88 Stat. 396, added subsec. (g).]

REPRESENTATIVES AND ELECTIONS

Sec. 9 [§ 159.]

(a) **[Exclusive representatives; employees' adjustment of grievances directly with employer]** Representatives designated or selected for the purposes of collective bargaining by the majority of the employees in a unit appropriate for such purposes, shall be the exclusive representatives of all the employees in such unit for the purposes of collective bargaining in respect to rates of pay, wages, hours of employment, or other conditions of employment: *Provided*, That any individual employee or a group of employees shall have the right at any time to present grievances to their employer and to have such grievances adjusted, without the intervention of the bargaining representative, as long as the adjustment is not inconsistent with the terms of a collective-bargaining contract or agreement then in effect: *Provided further*, That the bargaining representative has been given opportunity to be present at such adjustment.

(b) **[Determination of bargaining unit by Board]** The Board shall decide in each case whether, in order to assure to employees the fullest freedom in exercising the rights guaranteed by this Act [subchapter], the unit appropriate for the purposes of collective bargaining shall be the employer unit, craft unit, plant unit, or subdivision thereof: *Provided*, That the Board shall not (1) decide that any unit is appropriate for such purposes if such unit includes both professional employees and employees who are not professional employees unless a majority of such professional employees vote for inclusion in such unit; or (2) decide that any craft unit is inappropriate for such purposes on the ground that a different unit has been established by a prior Board determination, unless a majority of the employees in the proposed craft unit votes against separate representation or (3) decide that any unit is appropriate for such purposes if it includes, together with other employees, any individual employed as a guard to enforce against employees and other persons rules to protect property of the employer or to protect the safety of persons on the employer's premises; but no labor organization shall be certified as the representative of employees in a bargaining unit of guards if such organization admits to membership, or is affiliated directly or indirectly with an organization which admits to membership, employees other than guards.

(c) **[Hearings on questions affecting commerce; rules and regulations]**

(1) Whenever a petition shall have been filed, in accordance with such regulations as may be prescribed by the Board—

(A) by an employee or group of employees or any individual or labor organization acting in their behalf alleging that a substantial number of employees (i) wish to be represented for collective bargaining and that their employer declines to recognize their representative as the representative defined in section 9(a) [subsection (a) of this section], or (ii) assert that the individual or labor organization, which has been certified or is being currently recognized by their employer as the bargaining representative, is no longer a representative as defined in section 9(a) [subsection (a) of this section]; or

(B)   by an employer, alleging that one or more individuals or labor organizations have presented to him a claim to be recognized as the representative defined in section 9(a) [subsection (a) of this section]; the Board shall investigate such petition and if it has reasonable cause to believe that a question of representation affecting commerce exists shall provide for an appropriate hearing upon due notice. Such hearing may be conducted by an officer or employee of the regional office, who shall not make any recommendations with respect thereto. If the Board finds upon the record of such hearing that such a question of representation exists, it shall direct an election by secret ballot and shall certify the results thereof.

(2)   In determining whether or not a question of representation affecting commerce exists, the same regulations and rules of decision shall apply irrespective of the identity of the persons filing the petition or the kind of relief sought and in no case shall the Board deny a labor organization a place on the ballot by reason of an order with respect to such labor organization or its predecessor not issued in conformity with section 10(c) [section 160(c) of this title].

(3)   No election shall be directed in any bargaining unit or any subdivision within which, in the preceding twelve-month period, a valid election shall have been held. Employees engaged in an economic strike who are not entitled to reinstatement shall be eligible to vote under such regulations as the Board shall find are consistent with the purposes and provisions of this Act [subchapter] in any election conducted within twelve months after the commencement of the strike. In any election where none of the choices on the ballot receives a majority, a run-off shall be conducted, the ballot providing for a selection between the two choices receiving the largest and second largest number of valid votes cast in the election.

(4)   Nothing in this section shall be construed to prohibit the waiving of hearings by stipulation for the purpose of a consent election in conformity with regulations and rules of decision of the Board.

(5)   In determining whether a unit is appropriate for the purposes specified in subsection (b) [of this section] the extent to which the employees have organized shall not be controlling.

(d)   **[Petition for enforcement or review; transcript]** Whenever an order of the Board made pursuant to section 10(c) [section 160(c) of this title] is based in whole or in part upon facts certified following an investigation pursuant to subsection (c) of this section and there is a petition for the enforcement or review of such order, such certification and the record of such investigation shall be included in the transcript of the entire record required to be filed under section 10(e) or 10(f) [subsection (e) or (f) of section 160 of this title], and thereupon the decree of the court enforcing, modifying, or setting aside in whole or in part the order of the Board shall be made and entered upon the pleadings, testimony, and proceedings set forth in such transcript.

(e)   **[Secret ballot; limitation of elections]**

    (1) Upon the filing with the Board, by 30 per centum or more of the employees in a bargaining unit covered by an agreement between their employer and labor organization made pursuant to section 8(a)(3) [section 158(a)(3) of this title], of a petition alleging they desire that such authorization be rescinded, the Board shall take a secret ballot of the employees in such unit and certify the results thereof to such labor organization and to the employer.

    (2) No election shall be conducted pursuant to this subsection in any bargaining unit or any subdivision within which, in the preceding twelve-month period, a valid election shall have been held.

PREVENTION OF UNFAIR LABOR PRACTICES

Sec. 10. [§ 160.]

(a)   **[Powers of Board generally]** The Board is empowered, as hereinafter provided, to prevent any person from engaging in any unfair labor practice (listed in section 8 [section 158 of this title]) affecting commerce. This power shall not be affected by any other means of adjustment or prevention that has been or may be established by agreement, law, or otherwise: *Provided*, That the Board is empowered by agreement with any agency of any State or Territory to cede to such agency jurisdiction over any cases in any industry (other than mining, manufacturing, communications, and transportation except where predominantly local in character) even though such cases may involve labor disputes affecting commerce, unless the provision of the State or Territorial statute applicable to the determination of such cases by such agency is inconsistent with the corresponding provision of this Act [subchapter] or has received a construction inconsistent therewith.

(b)   **[Complaint and notice of hearing; six-month limitation; answer; court rules of evidence inapplicable]** Whenever it is charged that any person has engaged in or is engaging in any such unfair labor practice, the Board, or any agent or agency designated by the Board for such purposes, shall have power to issue and cause to be served upon such person a complaint stating the charges in that respect, and containing a notice of hearing before the Board or a member thereof, or before a designated agent or agency, at a place therein fixed, not less than five days after the serving of said complaint: *Provided*, That no complaint shall issue based upon any unfair labor practice occurring more than six months prior to the filing of the charge with the Board and the service of a copy thereof upon the person against whom such charge is made, unless the person aggrieved thereby was prevented from filing such charge by reason of service in the armed forces, in which event the six-month period shall be computed from the day of his discharge. Any such complaint may be amended by the member, agent, or agency conducting the hearing or the Board in its discretion at any time prior to the issuance of an order based thereon. The person so complained of shall have the right to file an answer to the original or amended complaint and to appear in person or otherwise and give testimony at the place and time fixed in the complaint. In the discretion of the member, agent, or agency conducting the hearing or the Board, any other person may be allowed to intervene in the said proceeding and to present testimony. Any such proceeding shall, so far as practicable, be conducted in accordance with the rules of evidence applicable in the district courts of the United States under the rules of civil procedure for the district courts of the United States, adopted by the Supreme Court of the United States pursuant to section 2072 of title 28, United States Code [section 2072 of title 28].

(c)   **[Reduction of testimony to writing; findings and orders of Board]**
The testimony taken by such member, agent, or agency, or the Board shall be reduced to writing and filed with the Board. Thereafter, in its discretion, the Board upon notice may take further testimony or hear argument. If upon the preponderance of the testimony taken the Board shall be of the opinion that any person named in the complaint has engaged in or is engaging in any such unfair labor practice, then the Board shall state its findings of fact and shall issue and cause to be served on such person an order requiring such person to cease and desist from such unfair labor practice, and to take such affirmative action including reinstatement of employees with or without backpay, as will effectuate the policies of this Act [subchapter]: *Provided*, That where an order directs reinstatement of an employee, backpay may be required of the employer or labor organization, as the case may be, responsible for the discrimination suffered by him: *And provided further*, That in determining whether a complaint shall issue alleging a violation of section 8(a)(1) or section 8(a)(2) [subsection (a)(1) or (a)(2) of section 158 of this title], and in deciding such cases, the same regulations and rules of decision shall apply irrespective of whether or not the labor organization affected is affiliated with a labor organization national or international in scope. Such order may further require such person to make reports from time to time showing the extent to which it has complied with the order. If upon the preponderance of the testimony taken the Board shall not be of the opinion that the person named in the complaint has engaged in or is engaging in any such unfair labor practice, then the Board shall state its findings of fact and shall issue an order dismissing the said complaint. No order of the Board shall require the reinstatement of any individual as an employee who has been suspended or discharged, or

the payment to him of any backpay, if such individual was suspended or discharged for cause. In case the evidence is presented before a member of the Board, or before an administrative law judge or judges thereof, such member, or such judge or judges, as the case may be, shall issue and cause to be served on the parties to the proceeding a proposed report, together with a recommended order, which shall be filed with the Board, and if no exceptions are filed within twenty days after service thereof upon such parties, or within such further period as the Board may authorize, such recommended order shall become the order of the Board and become affective as therein prescribed.

[The title "administrative law judge" was adopted in 5 U.S.C. § 3105.]

(d) **[Modification of findings or orders prior to filing record in court]** Until the record in a case shall have been filed in a court, as hereinafter provided, the Board may at any time, upon reasonable notice and in such manner as it shall deem proper, modify or set aside, in whole or in part, any finding or order made or issued by it.

(e) **[Petition to court for enforcement of order; proceedings; review of judgment]** The Board shall have power to petition any court of appeals of the United States, or if all the courts of appeals to which application may be made are in vacation, any district court of the United States, within any circuit or district, respectively, wherein the unfair labor practice in question occurred or wherein such person resides or transacts business, for the enforcement of such order and for appropriate temporary relief or restraining order, and shall file in the court the record in the proceeding, as provided in section 2112 of title 28, United States Code [section 2112 of title 28]. Upon the filing of such petition, the court shall cause notice thereof to be served upon such person, and thereupon shall have jurisdiction of the proceeding and of the question determined therein, and shall have power to grant such temporary relief or restraining order as it deems just and proper, and to make and enter a decree enforcing, modifying and enforcing as so modified, or setting aside in whole or in part the order of the Board. No objection that has not been urged before the Board, its member, agent, or agency, shall be considered by the court, unless the failure or neglect to urge such objection shall be excused because of extraordinary circumstances. The findings of the Board with respect to questions of fact if supported by substantial evidence on the record considered as a whole shall be conclusive. If either party shall apply to the court for leave to adduce additional evidence and shall show to the satisfaction of the court that such additional evidence is material and that there were reasonable grounds for the failure to adduce such evidence in the hearing before the Board, its member, agent, or agency, the court may order such additional evidence to be taken before the Board, its member, agent, or agency, and to be made a part of the record. The Board may modify its findings as to the facts, or make new findings, by reason of additional evidence so taken and filed, and it shall file such modified or new findings, which findings with respect to question of fact if supported by substantial evidence on the record considered as a whole shall be conclusive, and shall file its recommendations, if any, for the modification or setting aside of its original order. Upon the filing of the record with it the jurisdiction of the court shall be exclusive and its judgment and decree shall be final, except that the same shall be subject to review by the appropriate United States court of appeals if application was made to the district court as hereinabove provided, and by the Supreme Court of the United States upon writ of certiorari or certification as provided in section 1254 of title 28.

(f) **[Review of final order of Board on petition to court]** Any person aggrieved by a final order of the Board granting or denying in whole or in part the relief sought may obtain a review of such order in any United States court of appeals in the circuit wherein the unfair labor practice in question was alleged to have been engaged in or wherein such person resides or transacts business, or in the United States Court of Appeals for the District of Columbia, by filing in such court a written petition praying that the order of the Board be modified or set aside. A copy of such petition shall be forthwith transmitted by the clerk of the court to the Board, and thereupon the aggrieved party shall file in the court the record in the proceeding, certified by the Board, as provided in section 2112 of title 28, United States Code [section 2112 of title 28]. Upon the filing of such petition, the court shall proceed in the same manner as in the case of an application by the Board under subsection (e) of this section, and shall have the same jurisdiction to grant to the Board such temporary relief or restraining order as it deems just and proper, and in like manner to make and enter a decree enforcing, modifying and enforcing as so modified, or setting aside in whole or in part the order of the Board; the findings of the Board with respect to questions of fact if supported by substantial evidence on the record considered as a whole shall in like manner be conclusive.

(g) **[Institution of court proceedings as stay of Board's order]** The commencement of proceedings under subsection (e) or (f) of this section shall not, unless specifically ordered by the court, operate as a stay of the Board's order.

(h) **[Jurisdiction of courts unaffected by limitations prescribed in chapter 6 of this title]** When granting appropriate temporary relief or a restraining order, or making and entering a decree enforcing, modifying and enforcing as so modified, or setting aside in whole or in part an order of the Board, as provided in this section, the jurisdiction of courts sitting in equity shall not be limited by sections 101 to 115 of title 29, United States Code [chapter 6 of this title] [known as the "Norris-LaGuardia Act"].

(i) **Repealed.**

(j) **[Injunctions]** The Board shall have power, upon issuance of a complaint as provided in subsection (b) [of this section] charging that any person has engaged in or is engaging in an unfair labor practice, to petition any United States district court, within any district wherein the unfair labor practice in question is alleged to have occurred or wherein such person resides or transacts business, for appropriate temporary relief or restraining order. Upon the filing of any such petition the court shall cause notice thereof to be served upon such person, and thereupon shall have jurisdiction to grant to the Board such temporary relief or restraining order as it deems just and proper.

(k)   **[Hearings on jurisdictional strikes]** Whenever it is charged that any person has engaged in an unfair labor practice within the meaning of paragraph (4)(D) of section 8(b) [section 158(b) of this title], the Board is empowered and directed to hear and determine the dispute out of which such unfair labor practice shall have arisen, unless, within ten days after notice that such charge has been filed, the parties to such dispute submit to the Board satisfactory evidence that they have adjusted, or agreed upon methods for the voluntary adjustment of, the dispute. Upon compliance by the parties to the dispute with the decision of the Board or upon such voluntary adjustment of the dispute, such charge shall be dismissed.

(l)   **[Boycotts and strikes to force recognition of uncertified labor organizations; injunctions; notice; service of process]** Whenever it is charged that any person has engaged in an unfair labor practice within the meaning of paragraph (4)(A), (B), or (C) of section 8(b) [section 158(b) of this title], or section 8(e) [section 158(e) of this title] or section 8(b)(7) [section 158(b)(7) of this title], the preliminary investigation of such charge shall be made forthwith and given priority over all other cases except cases of like character in the office where it is filed or to which it is referred. If, after such investigation, the officer or regional attorney to whom the matter may be referred has reasonable cause to believe such charge is true and that a complaint should issue, he shall, on behalf of the Board, petition any United States district court within any district where the unfair labor practice in question has occurred, is alleged to have occurred, or wherein such person resides or transacts business, for appropriate injunctive relief pending the final adjudication of the Board with respect to such matter. Upon the filing of any such petition the district court shall have jurisdiction to grant such injunctive relief or temporary restraining order as it deems just and proper, notwithstanding any other provision of law: *Provided further,* That no temporary restraining order shall be issued without notice unless a petition alleges that substantial and irreparable injury to the charging party will be unavoidable and such temporary restraining order shall be effective for no longer than five days and will become void at the expiration of such period: *Provided further,* That such officer or regional attorney shall not apply for any restraining order under section 8(b)(7) [section 158(b)(7) of this title] if a charge against the employer under section 8(a)(2) [section 158(a)(2) of this title] has been filed and after the preliminary investigation, he has reasonable cause to believe that such charge is true and that a complaint should issue. Upon filing of any such petition the courts shall cause notice thereof to be served upon any person involved in the charge and such person, including the charging party, shall be given an opportunity to appear by counsel and present any relevant testimony: *Provided further,* That for the purposes of this subsection district courts shall be deemed to have jurisdiction of a labor organization (1) in the district in which such organization maintains its principal office, or (2) in any district in which its duly authorized officers or agents are engaged in promoting or protecting the interests of employee members. The service of legal process upon such officer or agent shall constitute service upon the labor organization and make such organization a party to the suit. In situations where such relief is appropriate the procedure specified herein shall apply to charges with respect to section 8(b)(4)(D) [section 158(b)(4)(D) of this title].

(m) **[Priority of cases]** Whenever it is charged that any person has engaged in an unfair labor practice within the meaning of subsection (a)(3) or (b)(2) of section 8 [section 158 of this title], such charge shall be given priority over all other cases except cases of like character in the office where it is filed or to which it is referred and cases given priority under subsection (1) [of this section].

INVESTIGATORY POWERS

Sec. 11. [§ 161.] For the purpose of all hearings and investigations, which, in the opinion of the Board, are necessary and proper for the exercise of the powers vested in it by section 9 and section 10 [sections 159 and 160 of this title]—

(1) **[Documentary evidence; summoning witnesses and taking testimony]** The Board, or its duly authorized agents or agencies, shall at all reasonable times have access to, for the purpose of examination, and the right to copy any evidence of any person being investigated or proceeded against that relates to any matter under investigation or in question. The Board, or any member thereof, shall upon application of any party to such proceedings, forthwith issue to such party subpoenas requiring the attendance and testimony of witnesses or the production of any evidence in such proceeding or investigation requested in such application. Within five days after the service of a subpoena on any person requiring the production of any evidence in his possession or under his control, such person may petition the Board to revoke, and the Board shall revoke, such subpoena if in its opinion the evidence whose production is required does not relate to any matter under investigation, or any matter in question in such proceedings, or if in its opinion such subpoena does not describe with sufficient particularity the evidence whose production is required. Any member of the Board, or any agent or agency designated by the Board for such purposes, may administer

oaths and affirmations, examine witnesses, and receive evidence. Such attendance of witnesses and the production of such evidence may be required from any place in the United States or any Territory or possession thereof, at any designated place of hearing.

(2) **[Court aid in compelling production of evidence and attendance of witnesses]** In case on contumacy or refusal to obey a subpoena issued to any person, any United States district court or the United States courts of any Territory or possession, within the jurisdiction of which the inquiry is carried on or within the jurisdiction of which said person guilty of contumacy or refusal to obey is found or resides or transacts business, upon application by the Board shall have jurisdiction to issue to such person an order requiring such person to appear before the Board, its member, agent, or agency, there to produce evidence if so ordered, or there to give testimony touching the matter under investigation or in question; and any failure to obey such order of the court may be punished by said court as a contempt thereof.

(3) **Repealed.**

[Immunity of witnesses. See 18 U.S.C. § 6001 et seq.]

(4) **[Process, service and return; fees of witnesses]** Complaints, orders and other process and papers of the Board, its member, agent, or agency, may be served either personally or by registered or certified mail or by telegraph or by leaving a copy thereof at the principal office or place of business of the person required to be served. The verified return by the individual so serving the same setting forth the manner of such service shall be proof of the same, and the return post office receipt or telegraph receipt therefore when registered or certified and mailed or when telegraphed as aforesaid shall be proof of service of the same. Witnesses summoned before the Board, its member, agent, or agency, shall be paid the same fees and mileage that are paid witnesses in the courts of the United States, and witnesses whose depositions are taken and the persons taking the same shall severally be entitled to the same fees as are paid for like services in the courts of the United States.

(5) **[Process, where served]** All process of any court to which application may be made under this Act [subchapter] may be served in the judicial district wherein the defendant or other person required to be served resides or may be found.

(6) **[Information and assistance from departments]** The several departments and agencies of the Government, when directed by the President, shall furnish the Board, upon its request, all records, papers, and information in their possession relating to any matter before the Board.

Sec. 12. [§ 162. Offenses and penalties] Any person who shall willfully resist, prevent, impede, or interfere with any member of the Board or any of its agents or agencies in the performance of duties pursuant to this Act [subchapter] shall be punished by a fine of not more than $5,000 or by imprisonment for not more than one year, or both.

LIMITATIONS

Sec. 13. [§ 163. Right to strike preserved] Nothing in this Act [subchapter], except as specifically provided for herein, shall be construed so as either to interfere with or impede or diminish in any way the right to strike or to affect the limitations or qualifications on that right.

Sec. 14. [§ 164. Construction of provisions]

(a) **[Supervisors as union members]** Nothing herein shall prohibit any individual employed as a supervisor from becoming or remaining a member of a labor organization, but no employer subject to this Act [subchapter] shall be compelled to deem individuals defined herein as supervisors as employees for the purpose of any law, either national or local, relating to collective bargaining.

(b) **[Agreements requiring union membership in violation of State law]** Nothing in this Act [subchapter] shall be construed as authorizing the execution or application of agreements requiring membership in a labor organization as a condition of employment in any State or Territory in which such execution or application is prohibited by State or Territorial law.

(c) **[Power of Board to decline jurisdiction of labor disputes; assertion of jurisdiction by State and Territorial courts]**

(1) The Board, in its discretion, may, by rule of decision or by published rules adopted pursuant to the Administrative Procedure Act [to subchapter II of chapter 5 of title 5], decline to assert jurisdiction over any labor dispute involving any class or category of employers, where, in the opinion of the Board, the effect of such labor dispute on commerce is not sufficiently substantial to warrant the exercise of its jurisdiction: *Provided,* That the Board shall not decline to assert jurisdiction over any labor dispute over which it would assert jurisdiction under the standards prevailing upon August 1, 1959.

(2) Nothing in this Act [subchapter] shall be deemed to prevent or bar any agency or the courts of any State or Territory (including the Commonwealth of Puerto Rico, Guam, and the Virgin Islands), from assuming and asserting jurisdiction over labor disputes over which the Board declines, pursuant to paragraph (1) of this subsection, to assert jurisdiction.

Sec. 15. [§ 165.] Omitted.

[Reference to repealed provisions of bankruptcy statute.]

Sec. 16. [§ 166. Separability of provisions] If any provision of this Act [subchapter], or the application of such provision to any person or circumstances, shall be held invalid, the remainder of this Act [subchapter], or the application of such provision to persons or circumstances other than those as to which it is held invalid, shall not be affected thereby.

Sec. 17. [§ 167. Short title] This Act [subchapter] may be cited as the "National Labor Relations Act."

Sec. 18. [§ 168.] Omitted.

[Reference to former sec. 9(f), (g), and (h).]

INDIVIDUALS WITH RELIGIOUS CONVICTIONS

Sec. 19. [§ 169.] Any employee who is a member of and adheres to established and traditional tenets or teachings of a bona fide religion, body, or sect which has historically held conscientious objections to joining or financially supporting labor organizations shall not be required to join or financially support any labor organization as a condition of employment; except that such employee may be required in a contract between such employee's employer and a labor organization in lieu of periodic dues and initiation fees, to pay sums equal to such dues and initiation fees to a nonreligious, nonlabor organization charitable fund exempt from taxation under section 501(c)(3) of title 26 of the Internal Revenue Code [section 501(c)(3) of title 26], chosen by such employee from a list of at least three such funds, designated in such contract or if the contract fails to designate such funds, then to any such fund chosen by the employee. If such employee who holds conscientious objections pursuant to this section requests the labor organization to use the grievance-arbitration procedure on the employee's behalf, the labor organization is authorized to charge the employee for the reasonable cost of using such procedure.

[Sec. added, Pub. L. 93-360, July 26, 1974, 88 Stat. 397, and amended, Pub. L. 96-593, Dec. 24, 1980, 94 Stat. 3452.]

LABOR MANAGEMENT RELATIONS ACT

Also cited LMRA; 29 U.S.C. §§ 141–197
[Title 29, Chapter 7, United States Code]

SHORT TITLE AND DECLARATION OF POLICY

Section 1. [§ 141.]

(a) This Act [chapter] may be cited as the "Labor Management Relations Act, 1947." [Also known as the "Taft-Hartley Act."]

(b) Industrial strife which interferes with the normal flow of commerce and with the full production of articles and commodities for commerce, can be avoided or substantially minimized if employers, employees, and labor organizations each recognize under law one another's legitimate rights in their relations with each other, and above all recognize under law that neither party has any right in its relations with any other to engage in acts or practices which jeopardize the public health, safety, or interest.

It is the purpose and policy of this Act [chapter], in order to promote the full flow of commerce, to prescribe the legitimate rights of both employees and employers in their relations affecting commerce, to provide orderly and peaceful procedures for preventing the interference by either with the legitimate rights of the other, to protect the rights of individual employees in their relations with labor organizations whose activities affect commerce, to define and proscribe practices on the part of labor and management which affect commerce and are inimical to the general welfare, and to protect the rights of the public in connection with labor disputes affecting commerce.

TITLE I, AMENDMENTS TO NATIONAL LABOR RELATIONS ACT

29 U.S.C. §§ 151–169 (printed above)

TITLE II

[Title 29, Chapter 7, Subchapter III, United States Code]

CONCILIATION OF LABOR DISPUTES IN INDUSTRIES AFFECTING COMMERCE; NATIONAL EMERGENCIES

Sec. 201. [§ 171. Declaration of purpose and policy] It is the policy of the United States that—

(a) sound and stable industrial peace and the advancement of the general welfare, health, and safety of the Nation and of the best interest of employers and employees can most satisfactorily be secured by the settlement of issues between employers and employees through the processes of conference and collective bargaining between employers and the representatives of their employees;

(b) the settlement of issues between employers and employees through collective bargaining may by advanced by making available full and adequate governmental facilities for conciliation, mediation, and voluntary arbitration to aid and encourage employers and the representatives of their employees to reach and maintain agreements concerning rates of pay, hours, and working conditions, and to make all reasonable efforts to settle their differences by mutual agreement reached through conferences and collective bargaining or by such methods as may be provided for in any applicable agreement for the settlement of disputes; and

(c) certain controversies which arise between parties to collective bargaining agreements may be avoided or minimized by making available full and adequate governmental facilities for furnishing assistance to employers and the representatives of their employees in formulating for inclusion within such agreements provision for adequate notice of any proposed changes in the terms of such agreements, for the final adjustment of grievances or questions regarding the application or interpretation of such agreements, and other provisions designed to prevent the subsequent arising of such controversies.

Sec. 202. [§ 172. Federal Mediation and Conciliation Service]

(a) **[Creation; appointment of Director]** There is created an independent agency to be known as the Federal Mediation and Conciliation Service (herein referred to as the "Service," except that for sixty days after June 23, 1947, such term shall refer to the Conciliation Service of the Department of Labor). The Service shall be under the direction of a Federal Mediation and Conciliation Director (hereinafter referred to as the "Director"), who shall be appointed by the President by and with the advice and consent of the Senate. The Director shall not engage in any other business, vocation, or employment.

(b) **[Appointment of officers and employees; expenditures for supplies, facilities, and services]** The Director is authorized, subject to the civil service laws, to appoint such clerical and other personnel as may be necessary for the execution of the functions of the Service, and shall fix their compensation in accordance with sections 5101 to 5115 and sections 5331 to 5338 of title 5, United States Code [chapter 51 and subchapter III of chapter 53 of title 5], and may, without regard to the provisions of the civil service laws, appoint such conciliators and mediators as may be necessary to carry out the functions of the Service. The Director is authorized to make such expenditures for supplies, facilities, and services as he deems necessary. Such expenditures shall be allowed and paid upon presentation of itemized vouchers therefore approved by the Director or by any employee designated by him for that purpose.

(c) **[Principal and regional offices; delegation of authority by Director; annual report to Congress]** The principal office of the Service shall be in the District of Columbia, but the Director may establish regional offices convenient to localities in which labor controversies are likely to arise. The Director may by order, subject to revocation at any time, delegate any authority and discretion conferred upon him by this Act [chapter] to any regional director, or other officer or employee of the Service. The Director may establish suitable procedures for cooperation with State and local mediation agencies. The Director shall make an annual report in writing to Congress at the end of the fiscal year.

(d) **[Transfer of all mediation and conciliation services to Service; effective date; pending proceedings unaffected]** All mediation and conciliation functions of the Secretary of Labor or the United States Conciliation Service under section 51 [repealed] of title 29, United States Code [this title], and all functions of the United States Conciliation Service under any other law are transferred to the Federal Mediation and Conciliation Service, together with the personnel and records of the United States Conciliation Service. Such transfer shall take effect upon the sixtieth day after June 23, 1947. Such transfer shall not affect any proceedings pending before the United States Conciliation Service or any certification, order, rule, or regulation theretofore made by it or by the Secretary of Labor. The Director and the Service shall not be subject in any way to the jurisdiction or authority of the Secretary of Labor or any official or division of the Department of Labor.

FUNCTIONS OF THE SERVICE

Sec. 203. [§ 173. Functions of Service]

(a) **[Settlement of disputes through conciliation and mediation]** It shall be the duty of the Service, in order to prevent or minimize interruptions of the free flow of commerce growing out of labor disputes, to assist parties to labor disputes in industries affecting commerce to settle such disputes through conciliation and mediation.

(b) **[Intervention on motion of Service or request of parties; avoidance of mediation of minor disputes]** The Service may proffer its services in any labor dispute in any industry affecting commerce, either upon its own motion or upon the request of one or more of the parties to the dispute, whenever in its judgment such dispute threatens to cause a substantial interruption of commerce. The Director and the Service are directed to avoid attempting to mediate disputes which would have only a minor effect on interstate commerce if State or other conciliation services are available to the parties. Whenever the Service does proffer its services in any dispute, it shall be the duty of the Service promptly to put itself in communication with the parties and to use its best efforts, by mediation and conciliation, to bring them to agreement.

(c) **[Settlement of disputes by other means upon failure of conciliation]** If the Director is not able to bring the parties to agreement by conciliation within a reasonable time, he shall seek to induce the parties voluntarily to seek other means of settling the dispute without resort to strike, lockout, or other coercion, including submission to the employees in the bargaining unit of the employer's last offer of settlement for approval or rejection in a secret ballot. The failure or refusal of either party to agree to any procedure suggested by the Director shall not be deemed a violation of any duty or obligation imposed by this Act [chapter].

(d) **[Use of conciliation and mediation services as last resort]** Final adjustment by a method agreed upon by the parties is declared to be the desirable method for settlement of grievance disputes arising over the application or interpretation of an existing collective-bargaining agreement. The Service is directed to make its conciliation and mediation services available in the settlement of such grievance disputes only as a last resort and in exceptional cases.

(e) **[Encouragement and support of establishment and operation of joint labor management activities conducted by committees]** The Service is authorized and directed to encourage and support the establishment and operation of joint labor management activities conducted by plant, area, and industry wide committees designed to improve labor management relationships, job security and organizational effectiveness, in accordance with the provisions of section 205A [section 175a of this title].

[Pub. L. 95-524, § 6(c)(1), Oct. 27, 1978, 92 Stat. 2020, added subsec. (e).]

Sec. 204. [§ 174. Co-equal obligations of employees, their representatives, and management to minimize labor disputes]

(a)   In order to prevent or minimize interruptions of the free flow of commerce growing out of labor disputes, employers and employees and their representatives, in any industry affecting commerce, shall—

(1)   exert every reasonable effort to make and maintain agreements concerning rates of pay, hours, and working conditions, including provision for adequate notice of any proposed change in the terms of such agreements;

(2)   whenever a dispute arises over the terms or application of a collective-bargaining agreement and a conference is requested by a party or prospective party thereto, arrange promptly for such a conference to be held and endeavor in such conference to settle such dispute expeditiously; and

(3)   in case such dispute is not settled by conference, participate fully and promptly in such meetings as may be undertaken by the Service under this Act [chapter] for the purpose of aiding in a settlement of the dispute.

Sec. 205. [§175. National Labor-Management Panel; creation and composition; appointment, tenure, and compensation; duties]

(a)   There is created a National Labor-Management Panel which shall be composed of twelve members appointed by the President, six of whom shall be elected from among persons outstanding in the field of management and six of whom shall be selected from among persons outstanding in the field of labor. Each member shall hold office for a term of three years, except that any member appointed to fill a vacancy occurring prior to the expiration of the term for which his predecessor was appointed shall be appointed for the remainder of such term, and the terms of office of the members first taking office shall expire, as designated by the President at the time of appointment, four at the end of the first year, four at the end of the second year, and four at the end of the third year after the date of appointment. Members of the panel, when serving on business of the panel, shall be paid compensation at the rate of $25 per day, and shall also be entitled to receive an allowance for actual and necessary travel and subsistence expenses while so serving away from their places of residence.

(b)   It shall be the duty of the panel, at the request of the Director, to advise in the avoidance of industrial controversies and the manner in which mediation and voluntary adjustment shall be administered, particularly with reference to controversies affecting the general welfare of the country.

Sec. 205A. [§ 175a. Assistance to plant, area, and industry wide labor management committees]

(a)   [Establishment and operation of plant, area, and industry wide committees]

(1) The Service is authorized and directed to provide assistance in the establishment and operation of plant, area and industry wide labor management committees which—

(A) have been organized jointly by employers and labor organizations representing employees in that plant, area, or industry; and

(B) are established for the purpose of improving labor management relationships, job security, organizational effectiveness, enhancing economic development or involving workers in decisions affecting their jobs including improving communication with respect to subjects of mutual interest and concern.

(2) The Service is authorized and directed to enter into contracts and to make grants, where necessary or appropriate, to fulfill its responsibilities under this section.

(b)   [Restrictions on grants, contracts, or other assistance]

(1) No grant may be made, no contract may be entered into and no other assistance may be provided under the provisions of this section to a plant labor management committee unless the employees in that plant are represented by a labor organization and there is in effect at that plant a collective bargaining agreement.

(2) No grant may be made, no contract may be entered into and no other assistance may be provided under the provisions of this section to an area or industry wide labor management committee unless its participants include any labor organizations certified or recognized as the representative of the employees of an employer participating in such committee. Nothing in this clause shall prohibit participation in an area or industry wide committee by an employer whose employees are not represented by a labor organization.

(3) No grant may be made under the provisions of this section to any labor management committee which the Service finds to have as one of its purposes the discouragement of the exercise of rights contained in section 7 of the National Labor Relations Act (29 U.S.C. § 157) [section 157 of this title], or the interference with collective bargaining in any plant, or industry.

(c)   [Establishment of office] The Service shall carry out the provisions of this section through an office established for that purpose.

(d)   [Authorization of appropriations] There are authorized to be appropriated to carry out the provisions of this section $10,000,000 for the fiscal year 1979, and such sums as may be necessary thereafter.

[Pub. L. 95-524, § 6(c)(2), Oct. 27, 1978, 92 Stat. 2020, added Sec. 205A.]

NATIONAL EMERGENCIES

Sec. 206. [§ 176. Appointment of board of inquiry by President; report; contents; filing with Service] Whenever in the opinion of the President of the United States, a threatened or actual strike or lockout affecting an entire industry or a substantial part thereof engaged in trade, commerce, transportation, transmission, or communication among the several States or with foreign nations, or engaged in the production of goods for commerce, will, if permitted to occur or to continue, imperil the national health or safety, he may appoint a board of inquiry to inquire into the issues involved in the dispute and to make a written report to him within such time as he shall prescribe. Such report shall include a statement of the facts with respect to the dispute, including each party's statement of its position but shall not contain any recommendations. The President shall file a copy of such report with the Service and shall make its contents available to the public.

Sec. 207. [§ 177. Board of inquiry]

  (a) **[Composition]** A board of inquiry shall be composed of a chairman and such other members as the President shall determine, and shall have power to sit and act in any place within the United States and to conduct such hearings either in public or in private, as it may deem necessary or proper, to ascertain the facts with respect to the causes and circumstances of the dispute.

  (b) **[Compensation]** Members of a board of inquiry shall receive compensation at the rate of $50 for each day actually spent by them in the work of the board, together with necessary travel and subsistence expenses.

  (c) **[Powers of discovery]** For the purpose of any hearing or inquiry conducted by any board appointed under this title, the provisions of sections 49 and 50 of title 15, United States Code [sections 49 and 50 of title 15] (relating to the attendance of witnesses and the production of books, papers, and documents) are made applicable to the powers and duties of such board.

Sec. 208. [§ 178. Injunctions during national emergency]

  (a) **[Petition to district court by Attorney General on direction of President]** Upon receiving a report from a board of inquiry the President may direct the Attorney General to petition any district court of the United States having jurisdiction of the parties to enjoin such strike or lockout or the continuing thereof, and if the court finds that such threatened or actual strike or lockout—

    (i) affects an entire industry or a substantial part thereof engaged in trade, commerce, transportation, transmission, or communication among the several States or with foreign nations, or engaged in the production of goods for commerce; and

    (ii) if permitted to occur or to continue, will imperil the national health or safety, it shall have jurisdiction to enjoin any such strike or lockout, or the continuing thereof, and to make such other orders as may be appropriate.

  (b) **[Inapplicability of chapter 6]** In any case, the provisions of sections 101 to 115 of title 29, United States Code [chapter 6 of this title] [known as the "Norris-LaGuardia Act"] shall not be applicable.

  (c) **[Review of orders]** The order or orders of the court shall be subject to review by the appropriate United States court of appeals and by the

Supreme Court upon writ of certiorari or certification as provided in section 1254 of title 28, United States Code [section 1254 of title 28].

Sec. 209. [§ 179. Injunctions during national emergency; adjustment efforts by parties during injunction period]

(a) **[Assistance of Service; acceptance of Service's proposed settlement]** Whenever a district court has issued an order under section 208 [section 178 of this title] enjoining acts or practices which imperil or threaten to imperil the national health or safety, it shall be the duty of the parties to the labor dispute giving rise to such order to make every effort to adjust and settle their differences, with the assistance of the Service created by this Act [chapter]. Neither party shall be under any duty to accept, in whole or in part, any proposal of settlement made by the Service.

(b) **[Reconvening of board of inquiry; report by board; contents; secret ballot of employees by National Labor Relations Board; certification of results to Attorney General]** Upon the issuance of such order, the President shall reconvene the board of inquiry which has previously reported with respect to the dispute. At the end of a sixty-day period (unless the dispute has been settled by that time), the board of inquiry shall report to the President the current position of the parties and the efforts which have been made for settlement, and shall include a statement by each party of its position and a statement of the employer's last offer of settlement. The President shall make such report available to the public. The National Labor Relations Board, within the succeeding fifteen days, shall take a secret ballot of the employees of each employer involved in the dispute on the question of whether they wish to accept the final offer of settlement made by their employer, as stated by him and shall certify the results thereof to the Attorney General within five days thereafter.

Sec. 210. [§ 180. Discharge of injunction upon certification of results of election or settlement; report to Congress] Upon the certification of the results of such ballot or upon a settlement being reached, whichever happens sooner, the Attorney General shall move the court to discharge the injunction, which motion shall then be granted and the injunction discharged. When such motion is granted, the President shall submit to the Congress a full and comprehensive report of the proceedings, including the findings of the board of inquiry and the ballot taken by the National Labor Relations Board, together with such recommendations as he may see fit to make for consideration and appropriate action.

COMPILATION OF COLLECTIVE-BARGAINING AGREEMENTS, ETC.

Sec. 211. [§ 181.]

(a) For the guidance and information of interested representatives of employers, employees, and the general public, the Bureau of Labor Statistics of the Department of Labor shall maintain a file of copies of all available collective bargaining agreements and other available agreements and actions thereunder settling or adjusting labor disputes. Such file shall be open to inspection under appropriate conditions prescribed by the Secretary of Labor, except that no specific information submitted in confidence shall be disclosed.

(b)   The Bureau of Labor Statistics in the Department of Labor is authorized to furnish upon request of the Service, or employers, employees, or their representatives, all available data and factual information which may aid in the settlement of any labor dispute, except that no specific information submitted in confidence shall be disclosed.

EXEMPTION OF RAILWAY LABOR ACT

Sec. 212. [§ 182.] The provisions of this title [subchapter] shall not be applicable with respect to any matter which is subject to the provisions of the Railway Labor Act [45 U.S.C. § 151 et seq.], as amended from time to time.

CONCILIATION OF LABOR DISPUTES IN THE HEALTH CARE INDUSTRY

Sec. 213. [§ 183.]

(a)   **[Establishment of Boards of Inquiry; membership]** If, in the opinion of the Director of the Federal Mediation and Conciliation Service, a threatened or actual strike or lockout affecting a health care institution will, if permitted to occur or to continue, substantially interrupt the delivery of health care in the locality concerned, the Director may further assist in the resolution of the impasse by establishing within 30 days after the notice to the Federal Mediation and Conciliation Service under clause (A) of the last sentence of section 8(d) [section 158(d) of this title] (which is required by clause (3) of such section 8(d) [section 158(d) of this title]), or within 10 days after the notice under clause (B), an impartial Board of Inquiry to investigate the issues involved in the dispute and to make a written report thereon to the parties within fifteen (15) days after the establishment of such a Board. The written report shall contain the findings of fact together with the Board's recommendations for settling the dispute, with the objective of achieving a prompt, peaceful and just settlement of the dispute. Each such Board shall be composed of such number of individuals as the Director may deem desirable. No member appointed under this section shall have any interest or involvement in the health care institutions or the employee organizations involved in the dispute.

(b)   **[Compensation of members of Boards of Inquiry]**

(1)   Members of any board established under this section who are otherwise employed by the Federal Government shall serve without compensation but shall be reimbursed for travel, subsistence, and other necessary expenses incurred by them in carrying out its duties under this section.

(2) Members of any board established under this section who are not subject to paragraph (1) shall receive compensation at a rate prescribed by the Director but not to exceed the daily rate prescribed for GS-18 of the General Schedule under section 5332 of title 5, United States Code [section 5332 of title 5], including travel for each day they are engaged in the performance of their duties under this section and shall be entitled to reimbursement for travel, subsistence, and other necessary expenses incurred by them in carrying out their duties under this section.

(c) **[Maintenance of status quo]** After the establishment of a board under subsection (a) of this section and for 15 days after any such board has issued its report, no change in the status quo in effect prior to the expiration of the contract in the case of negotiations for a contract renewal, or in effect prior to the time of the impasse in the case of an initial bargaining negotiation, except by agreement, shall be made by the parties to the controversy.

(d) **[Authorization of appropriations]** There are authorized to be appropriated such sums as may be necessary to carry out the provisions of this section.

TITLE III

[Title 29, Chapter 7, Subchapter IV, United States Code]

SUITS BY AND AGAINST LABOR ORGANIZATIONS

Sec. 301. [§ 185.]

(a) **[Venue, amount, and citizenship]** Suits for violation of contracts between an employer and a labor organization representing employees in an industry affecting commerce as defined in this Act [chapter], or between any such labor organization, may be brought in any district court of the United States having jurisdiction of the parties, without respect to the amount in controversy or without regard to the citizenship of the parties.

(b) **[Responsibility for acts of agent; entity for purposes of suit; enforcement of money judgments]** Any labor organization which represents employees in an industry affecting commerce as defined in this Act [chapter] and any employer whose activities affect commerce as defined in this Act [chapter] shall be bound by the acts of its agents. Any such labor organization may sue or be sued as an entity and in behalf of the employees whom it represents in the courts of the United States. Any money judgment against a labor organization in a district court of the United States shall be enforceable only against the organization as an entity and against its assets, and shall not be enforceable against any individual member or his assets.

(c) **[Jurisdiction]** For the purposes of actions and proceedings by or against labor organizations in the district courts of the United States, district courts shall be deemed to have jurisdiction of a labor organization (1) in the district in which such organization maintains its principal offices, or (2) in any district in which its duly authorized officers or agents are engaged in representing or acting for employee members.

(d) **[Service of process]** The service of summons, subpoena, or other legal process of any court of the United States upon an officer or agent of a labor organization, in his capacity as such, shall constitute service upon the labor organization.

(e) **[Determination of question of agency]** For the purposes of this section, in determining whether any person is acting as an "agent" of another person so as to make such other person responsible for his acts, the question of whether the specific acts performed were actually authorized or subsequently ratified shall not be controlling.

RESTRICTIONS ON PAYMENTS TO EMPLOYEE REPRESENTATIVES

Sec. 302. [§ 186.]

(a) **[Payment or lending, etc., of money by employer or agent to employees, representatives, or labor organizations]** It shall be unlawful for any employer or association of employers or any person who acts as a labor relations expert, adviser, or consultant to an employer or who acts in the interest of an employer to pay, lend, or deliver, or agree to pay, lend, or deliver, any money or other thing of value—

(1) to any representative of any of his employees who are employed in an industry affecting commerce; or

(2) to any labor organization, or any officer or employee thereof, which represents, seeks to represent, or would admit to membership, any of the employees of such employer who are employed in an industry affecting commerce;

(3) to any employee or group or committee of employees of such employer employed in an industry affecting commerce in excess of their normal compensation for the purpose of causing such employee or group or committee directly or indirectly to influence any other employees in the exercise of the right to organize and bargain collectively through representatives of their own choosing; or

(4) to any officer or employee of a labor organization engaged in an industry affecting commerce with intent to influence him in respect to any of his actions, decisions, or duties as a representative of employees or as such officer or employee of such labor organization.

(b) **[Request, demand, etc., for money or other thing of value]**

(1) It shall be unlawful for any person to request, demand, receive, or accept, or agree to receive or accept, any payment, loan, or delivery of any money or other thing of value prohibited by subsection (a) [of this section].

(2) It shall be unlawful for any labor organization, or for any person acting as an officer, agent, representative, or employee of such labor organization, to demand or accept from the operator of any motor vehicle (as defined in part II of the Interstate Commerce Act [49 U.S.C. § 301 et seq.]) employed in the transportation of property in commerce, or the employer of any such operator, any money or other thing of value payable to such organization or to an officer, agent, representative or employee thereof as a fee or charge for the unloading, or in connection with the unloading, of the cargo of such vehicle: *Provided*, That nothing in this paragraph shall be construed to make unlawful any payment by an employer to any of his employees as compensation for their services as employees.

(c) **[Exceptions]** The provisions of this section shall not be applicable (1) in respect to any money or other thing of value payable by an employer to any of his employees whose established duties include acting openly for such employer in matters of labor relations or personnel administration or to any representative of his employees, or to any officer or employee of a labor organization, who is also an employee or former employee of such employer, as compensation for, or by reason of, his service as an employee of such employer; (2) with respect to the payment or delivery of any money or other thing of value in satisfaction of a judgment of any court or a decision or award of an arbitrator or impartial chairman or in compromise, adjustment, settlement, or release of any claim, complaint, grievance, or dispute in the absence of fraud or duress; (3) with respect to the sale or purchase of an article or commodity at the prevailing market price in the regular course of business; (4) with respect to money deducted from the wages of employees in payment of membership dues in a labor organization: *Provided*, That the employer has received from each employee, on whose account such deductions are made, a written assignment which shall not be irrevocable for a period of more than one year, or beyond the termination date of the applicable collective agreement, whichever occurs sooner; (5) with respect to money or other thing of value paid to a trust fund established by such representative, for the sole and exclusive benefit of the employees of such employer, and their families and dependents (or of such employees, families, and dependents jointly with the employees of other employers making similar payments, and their families and dependents): *Provided*, That (A) such payments are held in trust for the purpose of paying, either from principal or income or both, for the benefit of employees, their families and dependents, for medical or hospital care, pensions on retirement or death of employees, compensation for injuries or illness resulting from occupational activity or insurance to provide any of the foregoing, or unemployment benefits or life insurance, disability and sickness insurance, or accident insurance; (B) the detailed basis on which such payments are to be made is specified in a written agreement with the employer, and employees and employers are equally represented in the administration of such fund, together with such neutral persons as the representatives of the employers and the representatives of employees may agree upon and in the event the employer and employee groups deadlock on the administration of such fund and there are no neutral persons empowered to break such deadlock, such agreement provides that the two groups shall agree on an impartial umpire to decide such dispute, or in event of their failure to agree within a reasonable length of time, an impartial umpire to decide such dispute shall, on petition of either group, be appointed by the district court of the United States for the district where the trust fund has its principal office, and shall also contain provisions for an annual audit of the trust fund, a statement of the results of which shall be available for inspection by interested persons at the principal office of the trust fund and at such other places as may be designated in such written agreement; and (C) such payments as are intended to be used for the purpose of providing pensions or annuities for employees are made to a separate trust which provides that the funds held therein cannot be used for any purpose other than paying such pensions or annuities; (6) with respect to money or other thing of value paid by any employer to a trust fund established by such representative for the purpose of pooled vacation, holiday, severance or similar benefits, or defraying costs of apprenticeship or other training programs: *Provided*, That the requirements of clause (B) of the proviso to clause (5) of this subsection shall apply to such trust funds; (7) with respect to money or other thing of value paid by any employer to a pooled or individual trust fund established by such representative for the purpose of (A) scholarships for the benefit of employees, their families, and dependents for study at educational institutions, (B) child care centers for preschool and school age dependents of employees, or (C) financial assistance for employee housing: *Provided*, That no labor organization or employer shall be required to bargain on the establishment of any such trust fund, and refusal to do so shall not constitute an unfair labor practice: *Provided further*, That the requirements of clause (B) of the proviso to clause (5) of this subsection shall apply to such trust funds; (8) with respect to money or any other thing of value paid by any employer to a trust fund established by such representative for the purpose of defraying the costs of legal services for employees, their families, and dependents for counsel or plan of their choice: *Provided*, That the requirements of clause (B) of the proviso to clause (5) of this subsection shall apply to such trust funds: *Provided further*, That no such legal services shall be furnished: (A) to initiate any proceeding directed (i) against any such employer or its officers or agents except in workman's compensation cases, or (ii) against such labor organization, or its parent or subordinate bodies, or their officers or agents, or (iii) against any other employer or labor organization, or their officers or agents, in any matter arising under the National Labor Relations Act, or this Act [under subchapter II of this chapter or this chapter]; and (B) in any proceeding where a labor organization would be prohibited from defraying the costs of legal services by the provisions of the Labor-Management Reporting and Disclosure Act of 1959 [29 U.S.C. § 401 et seq.]; or (9) with respect to money or other things of value paid by an employer to a plant, area or industry wide labor management committee established for one or more of the purposes set forth in section 5(b) of the Labor Management Cooperation Act of 1978.

[Sec. 302(c)(7) was added by Pub. L. 91-86, Oct. 14, 1969, 83 Stat. 133; Sec. 302(c)(8) by Pub. L. 93-95, Aug. 15, 1973, 87 Stat. 314; Sec. 302(c)(9) by Pub. L. 95-524, Oct. 27, 1978, 92 Stat. 2021; and Sec. 302(c) (7) was amended by Pub. L. 101-273, Apr. 18, 1990, 104 Stat. 138.]

(d)   **[Penalty for violations]** Any person who willfully violates any of the provisions of this section shall, upon conviction thereof, be guilty of a misdemeanor and be subject to a fine of not more than $10,000 or to imprisonment for not more than one year, or both.

(e)   **[Jurisdiction of courts]** The district courts of the United States and the United States courts of the Territories and possessions shall have jurisdiction, for cause shown, and subject to the provisions of rule 65 of the Federal Rules of Civil Procedure [section 381 (repealed) of title 28] (relating to notice to opposite party) to restrain violations of this section, without regard to the provisions of section 7 of title 15 and section 52 of title 29, United States Code [of this title] [known as the "Clayton Act"], and the provisions of sections 101 to 115 of title 29, United States Code [chapter 6 of this title] [known as the "Norris-LaGuardia Act"].

(f)   **[Effective date of provisions]** This section shall not apply to any contract in force on June 23, 1947, until the expiration of such contract, or until July 1, 1948, whichever first occurs.

(g)   **[Contributions to trust funds]** Compliance with the restrictions contained in subsection (c)(5)(B) [of this section] upon contributions to trust funds, otherwise lawful, shall not be applicable to contributions to such trust funds established by collective agreement prior to January 1, 1946, nor shall subsection (c)(5)(A) [of this section] be construed as prohibiting contributions to such trust funds if prior to January 1, 1947, such funds contained provisions for pooled vacation benefits.

BOYCOTTS AND OTHER UNLAWFUL COMBINATIONS

Sec. 303. [§ 187.]

(a)   It shall be unlawful, for the purpose of this section only, in an industry or activity affecting commerce, for any labor organization to engage in any activity or conduct defined as an unfair labor practice in section 8(b) (4) of the National Labor Relations Act [section 158(b)(4) of this title].

(b)   Whoever shall be injured in his business or property by reason of any violation of subsection (a) [of this section] may sue therefore in any district court of the United States subject to the limitation and provisions of section 301 hereof [section 185 of this title] without respect to the amount in controversy, or in any other court having jurisdiction of the parties, and shall recover the damages by him sustained and the cost of the suit.

RESTRICTION ON POLITICAL CONTRIBUTIONS

Sec. 304. Repealed.
[See sec. 316 of the Federal Election Campaign Act of 1972, 2 U.S.C. § 441b.]

Sec. 305. [§ 188.] Strikes by Government employees. Repealed.
[See 5 U.S.C. § 7311 and 18 U.S.C. § 1918.]

TITLE IV

[Title 29, Chapter 7, Subchapter V, United States Code]

CREATION OF JOINT COMMITTEE TO STUDY AND REPORT ON BASIC PROBLEMS AFFECTING FRIENDLY LABOR RELATIONS AND PRODUCTIVITY

Secs. 401–407. [§§ 191–197.] Omitted.

TITLE V

[Title 29, Chapter 7, Subchapter I, United States Code]

DEFINITIONS

Sec. 501. [§ 142.] When used in this Act [chapter]—

(1) The term "industry affecting commerce" means any industry or activity in commerce or in which a labor dispute would burden or obstruct commerce or tend to burden or obstruct commerce or the free flow of commerce.

(2) The term "strike" includes any strike or other concerted stoppage of work by employees (including a stoppage by reason of the expiration of a collective-bargaining agreement) and any concerted slowdown or other concerted interruption of operations by employees.

(3) The terms "commerce," "labor disputes," "employer," "employee," "labor organization," "representative," "person," and "supervisor" shall have the same meaning as when used in the National Labor Relations Act as amended by this Act [in subchapter II of this chapter].

SAVING PROVISION

Sec. 502. [§ 143.] [Abnormally dangerous conditions] Nothing in this Act [chapter] shall be construed to require an individual employee to render labor or service without his consent, nor shall anything in this Act [chapter] be construed to make the quitting of his labor by an individual employee an illegal act; nor shall any court issue any process to compel the performance by an individual employee of such labor or service, without his consent; nor shall the quitting of labor by an employee or employees in good faith because of abnormally dangerous conditions for work at the place of employment of such employee or employees be deemed a strike under this Act [chapter].

SEPARABILITY

Sec. 503. [§ 144.] If any provision of this Act [chapter], or the application of such provision to any person or circumstance, shall be held invalid, the remainder of this Act [chapter], or the application of such provision to persons or circumstances other than those as to which it is held invalid, shall not be affected thereby.

Appendix 2

For historical purposes, this is the original text of the law, without any subsequent amendments. For the current texts of the laws we enforce, as amended, see Laws Enforced by the EEOC.

AN ACT

To amend the Civil Rights Act of 1964 to strengthen and improve Federal civil rights laws, to provide for damages in cases of intentional employment discrimination, to clarify provisions regarding disparate impact actions, and for other purposes.

Be it enacted by the Senate and House of Representatives of the United States of America in Congress assembled,

SECTION 1. SHORT TITLE.

This Act may be cited as the 'Civil Rights Act of 1991'.

SEC. 2. FINDINGS.

The Congress finds that—

(1) additional remedies under Federal law are needed to deter unlawful harassment and intentional discrimination in the workplace;
(2) the decision of the Supreme Court in Wards Cove Packing Co. v. Atonio, 490 U.S. 642 (1989) has weakened the scope and effectiveness of Federal civil rights protections; and
(3) legislation is necessary to provide additional protections against unlawful discrimination in employment.

SEC. 3. PURPOSES.

The purposes of this Act are—

(1) to provide appropriate remedies for intentional discrimination and unlawful harassment in the workplace;
(2) to codify the concepts of 'business necessity' and 'job related' enunciated by the Supreme Court in Griggs v. Duke Power Co., 401 U.S. 424 (1971), and in the other Supreme Court decisions prior to Wards Cove Packing Co. v. Atonio, 490 U.S. 642 (1989);

(3) to confirm statutory authority and provide statutory guidelines for the adjudication of disparate impact suits under title VII of the Civil Rights Act of 1964 (42 U.S.C. 2000e et seq.); and

(4) to respond to recent decisions of the Supreme Court by expanding the scope of relevant civil rights statutes in order to provide adequate protection to victims of discrimination.

TITLE I—FEDERAL CIVIL RIGHTS REMEDIES

Sec. 101. Prohibition Against All Racial Discrimination in the Making and Enforcement of Contracts.

Section 1977 of the Revised Statutes (42 U.S.C. 1981) is amended—

(1) by inserting '(a)' before 'All persons within'; and
(2) by adding at the end the following new subsections:
'(b)   For purposes of this section, the term 'make and enforce contracts' includes the making, performance, modification, and termination of contracts, and the enjoyment of all benefits, privileges, terms, and conditions of the contractual relationship.
'(c)   The rights protected by this section are protected against impairment by nongovernmental discrimination and impairment under color of State law.'

Sec. 102. Damages in Cases of Intentional Discrimination.

The Revised Statutes are amended by inserting after section 1977 (42 U.S.C. 1981) the following new section:

Sec. 1977A. Damages in Cases of Intentional Discrimination i Employment.

'(a) RIGHT OF RECOVERY—
'(1)   CIVIL RIGHTS—In an action brought by a complaining party under section 706 or 717 of the Civil Rights Act of 1964 (42 U.S.C. 2000e-5) against a respondent who engaged in unlawful intentional discrimination (not an employment practice that is unlawful because of its disparate impact) prohibited under section 703, 704, or 717 of the Act (42 U.S.C. 2000e-2 or 2000e-3), and provided that the complaining party cannot recover under section 1977 of the Revised Statutes (42 U.S.C. 1981), the complaining party may recover compensatory and punitive damages as allowed in subsection (b), in addition to any relief authorized by section 706(g) of the Civil Rights Act of 1964, from the respondent.
'(2)   DISABILITY—In an action brought by a complaining party under the powers, remedies, and procedures set forth in section 706 or 717 of the Civil Rights Act of 1964 (as provided in section 107(a) of the

Americans with Disabilities Act of 1990 (42 U.S.C. 12117(a)), and section 505(a)(1) of the Rehabilitation Act of 1973 (29 U.S.C. 794a(a)(1)), respectively) against a respondent who engaged in unlawful intentional discrimination (not an employment practice that is unlawful because of its disparate impact) under section 501 of the Rehabilitation Act of 1973 (29 U.S.C. 791) and the regulations implementing section 501, or who violated the requirements of section 501 of the Act or the regulations implementing section 501 concerning the provision of a reasonable accommodation, or section 102 of the Americans with Disabilities Act of 1990 (42 U.S.C. 12112), or committed a violation of section 102(b)(5) of the Act, against an individual, the complaining party may recover compensatory and punitive damages as allowed in subsection (b), in addition to any relief authorized by section 706(g) of the Civil Rights Act of 1964, from the respondent.

'(3) REASONABLE ACCOMMODATION AND GOOD FAITH EFFORT—In cases where a discriminatory practice involves the provision of a reasonable accommodation pursuant to section 102(b)(5) of the Americans with Disabilities Act of 1990 or regulations implementing section 501 of the Rehabilitation Act of 1973, damages may not be awarded under this section where the covered entity demonstrates good faith efforts, in consultation with the person with the disability who has informed the covered entity that accommodation is needed, to identify and make a reasonable accommodation that would provide such individual with an equally effective opportunity and would not cause an undue hardship on the operation of the business.

'(b) COMPENSATORY AND PUNITIVE DAMAGES—

'(1) DETERMINATION OF PUNITIVE DAMAGES—A complaining party may recover punitive damages under this section against a respondent (other than a government, government agency or political subdivision) if the complaining party demonstrates that the respondent engaged in a discriminatory practice or discriminatory practices with malice or with reckless indifference to the federally protected rights of an aggrieved individual.

'(2) EXCLUSIONS FROM COMPENSATORY DAMAGES— Compensatory damages awarded under this section shall not include backpay, interest on backpay, or any other type of relief authorized under section 706(g) of the Civil Rights Act of 1964.

'(3) LIMITATIONS—The sum of the amount of compensatory damages awarded under this section for future pecuniary losses, emotional pain, suffering, inconvenience, mental anguish, loss of enjoyment of life, and other nonpecuniary losses, and the amount of punitive damages awarded under this section, shall not exceed, for each complaining party—

'(A) in the case of a respondent who has more than 14 and fewer than 101 employees in each of 20 or more calendar weeks in the current or preceding calendar year, $50,000;

‘(B)   in the case of a respondent who has more than 100 and fewer than 201 employees in each of 20 or more calendar weeks in the current or preceding calendar year, $100,000; and

‘(C)   in the case of a respondent who has more than 200 and fewer than 501 employees in each of 20 or more calendar weeks in the current or preceding calendar year, $200,000; and

‘(D)   in the case of a respondent who has more than 500 employees in each of 20 or more calendar weeks in the current or preceding calendar year, $300,000.

‘(4) CONSTRUCTION—Nothing in this section shall be construed to limit the scope of, or the relief available under, section 1977 of the Revised Statutes (42 U.S.C. 1981).

‘(c) JURY TRIAL—If a complaining party seeks compensatory or punitive damages under this section—

‘(1) any party may demand a trial by jury; and

‘(2) the court shall not inform the jury of the limitations described in subsection (b)(3).

‘(d) DEFINITIONS—As used in this section:

‘(1) COMPLAINING PARTY—The term ‘complaining party’ means—

‘(A)   in the case of a person seeking to bring an action under subsection (a)(1), the Equal Employment Opportunity Commission, the Attorney General, or a person who may bring an action or proceeding under title VII of the Civil Rights Act of 1964 (42 U.S.C. 2000e et seq.); or

‘(B)   in the case of a person seeking to bring an action under subsection (a)(2), the Equal Employment Opportunity Commission, the Attorney General, a person who may bring an action or proceeding under section 505(a)(1) of the Rehabilitation Act of 1973 (29 U.S.C. 794a(a)(1)), or a person who may bring an action or proceeding under title I of the Americans with Disabilities Act of 1990 (42 U.S.C. 12101 et seq.).

‘(2) DISCRIMINATORY PRACTICE—The term ‘discriminatory practice’ means the discrimination described in paragraph (1), or the discrimination or the violation described in paragraph (2), of subsection (a).

Sec. 103. Attorney's Fees.

The last sentence of section 722 of the Revised Statutes (42 U.S.C. 1988) is amended by inserting ‘, 1977A’ after ‘1977’.

Sec. 104. Definitions.

Section 701 of the Civil Rights Act of 1964 (42 U.S.C. 2000e) is amended by adding at the end the following new subsections:

'(l) The term 'complaining party' means the Commission, the Attorney General, or a person who may bring an action or proceeding under this title.

'(m) The term 'demonstrates' means meets the burdens of production and persuasion.

'(n) The term 'respondent' means an employer, employment agency, labor organization, joint labor-management committee controlling apprenticeship or other training or retraining program, including an on-the-job training program, or Federal entity subject to section 717.'

Sec. 105. Burden of Proof in Disparate Impact Cases.

(a) Section 703 of the Civil Rights Act of 1964 (42 U.S.C. 2000e-2) is amended by adding at the end the following new subsection:

'(k)(1)(A) An unlawful employment practice based on disparate impact is established under this title only if—

'(i) a complaining party demonstrates that a respondent uses a particular employment practice that causes a disparate impact on the basis of race, color, religion, sex, or national origin and the respondent fails to demonstrate that the challenged practice is job related for the position in question and consistent with business necessity; or

'(ii) the complaining party makes the demonstration described in subparagraph (C) with respect to an alternative employment practice and the respondent refuses to adopt such alternative employment practice.

'(B)(i) With respect to demonstrating that a particular employment practice causes a disparate impact as described in subparagraph (A)(i), the complaining party shall demonstrate that each particular challenged employment practice causes a disparate impact, except that if the complaining party can demonstrate to the court that the elements of a respondent's decisionmaking process are not capable of separation for analysis, the decisionmaking process may be analyzed as one employment practice.

'(ii) If the respondent demonstrates that a specific employment practice does not cause the disparate impact, the respondent shall not be required to demonstrate that such practice is required by business necessity.

'(C) The demonstration referred to by subparagraph (A)(ii) shall be in accordance with the law as it existed on June 4, 1989, with respect to the concept of 'alternative employment practice'.

'(2) A demonstration that an employment practice is required by business necessity may not be used as a defense against a claim of intentional discrimination under this title.

'(3) Notwithstanding any other provision of this title, a rule barring the employment of an individual who currently and knowingly uses or possesses a controlled substance, as defined in schedules I and II of section 102(6) of the Controlled Substances Act (21 U.S.C. 802(6)), other than the use or possession of a drug taken under the supervision of a licensed health care professional, or any other use or possession authorized by the Controlled Substances Act or any other provision of Federal law, shall be considered an unlawful employment practice under this title only if such rule is adopted or applied with an intent to discriminate because of race, color, religion, sex, or national origin.'

(b) No statements other than the interpretive memorandum appearing at Vol. 137 Congressional Record S 15276 (daily ed. Oct. 25, 1991) shall be considered legislative history of, or relied upon in any way as legislative history in construing or applying, any provision of this Act that relates to Wards Cove—Business necessity/cumulation/alternative business practice.

SEC. 106. PROHIBITION AGAINST DISCRIMINATORY USE OF TEST SCORES.

Section 703 of the Civil Rights Act of 1964 (42 U.S.C. 2000e-2) (as amended by section 105) is further amended by adding at the end the following new subsection:

'(l) It shall be an unlawful employment practice for a respondent, in connection with the selection or referral of applicants or candidates for employment or promotion, to adjust the scores of, use different cutoff scores for, or otherwise alter the results of, employment related tests on the basis of race, color, religion, sex, or national origin.'

SEC. 107. CLARIFYING PROHIBITION AGAINST IMPERMISSIBLE CONSIDERATION OF RACE, COLOR, RELIGION, SEX, OR NATIONAL ORIGIN IN EMPLOYMENT PRACTICES.

(a) IN GENERAL—Section 703 of the Civil Rights Act of 1964 (42 U.S.C. 2000e-2) (as amended by sections 105 and 106) is further amended by adding at the end the following new subsection:

'(m) Except as otherwise provided in this title, an unlawful employment practice is established when the complaining party demonstrates that race, color, religion, sex, or national origin was a motivating factor for any employment practice, even though other factors also motivated the practice.'

(b) ENFORCEMENT PROVISIONS—Section 706(g) of such Act (42 U.S.C. 2000e-5(g)) is amended—

(1) by designating the first through third sentences as paragraph (1);

(2) by designating the fourth sentence as paragraph (2)(A) and indenting accordingly; and

(3) by adding at the end the following new subparagraph:

'(B) On a claim in which an individual proves a violation under section 703(m) and a respondent demonstrates that the respondent would have taken the same action in the absence of the impermissible motivating factor, the court—

'(i)   may grant declaratory relief, injunctive relief (except as provided in clause (ii)), and attorney's fees and costs demonstrated
to be directly attributable only to the pursuit of a claim under
section 703(m); and

'(ii)   shall not award damages or issue an order requiring any admis-
sion, reinstatement, hiring, promotion, or payment, described
in subparagraph (A).'

SEC. 108. FACILITATING PROMPT AND ORDERLY RESOLUTION OF CHALLENGES TO EMPLOYMENT PRACTICES IMPLEMENTING LITIGATED OR CONSENT JUDGMENTS OR ORDERS.

Section 703 of the Civil Rights Act of 1964 (42 U.S.C. 2000e-2) (as amended by
sections 105, 106, and 107 of this title) is further amended by adding at the end the
following new subsection:

'(n)(1)(A) Notwithstanding any other provision of law, and except as provided in
paragraph (2), an employment practice that implements and is within the
scope of a litigated or consent judgment or order that resolves a claim of
employment discrimination under the Constitution or Federal civil rights
laws may not be challenged under the circumstances described in subparagraph (B).

'(B)   A practice described in subparagraph (A) may not be challenged in
a claim under the Constitution or Federal civil rights laws—

'(i)   by a person who, prior to the entry of the judgment or order
described in subparagraph (A), had—

'(I)   actual notice of the proposed judgment or order sufficient
to apprise such person that such judgment or order might
adversely affect the interests and legal rights of such person
and that an opportunity was available to present objections to
such judgment or order by a future date certain; and

'(II)   a reasonable opportunity to present objections to such judgment or order; or

'(ii)   by a person whose interests were adequately represented by
another person who had previously challenged the judgment or
order on the same legal grounds and with a similar factual situation, unless there has been an intervening change in law or fact.

'(2)   Nothing in this subsection shall be construed to—

'(A)   alter the standards for intervention under rule 24 of the Federal
Rules of Civil Procedure or apply to the rights of parties who have
successfully intervened pursuant to such rule in the proceeding in
which the parties intervened;

'(B)   apply to the rights of parties to the action in which a litigated or consent judgment or order was entered, or of members of a class represented or sought to be represented in such action, or of members

of a group on whose behalf relief was sought in such action by the
Federal Government;

'(C)   prevent challenges to a litigated or consent judgment or order on the
       ground that such judgment or order was obtained through collusion
       or fraud, or is transparently invalid or was entered by a court lack-
       ing subject matter jurisdiction; or

'(D)   authorize or permit the denial to any person of the due process of
       law required by the Constitution.

'(3) Any action not precluded under this subsection that challenges an employment consent judgment or order described in paragraph (1) shall be brought
in the court, and if possible before the judge, that entered such judgment
or order. Nothing in this subsection shall preclude a transfer of such action
pursuant to section 1404 of title 28, United States Code.'

SEC. 109. PROTECTION OF EXTRATERRITORIAL EMPLOYMENT.

(a) DEFINITION OF EMPLOYEE—Section 701(f) of the Civil Rights Act
     of 1964 (42 U.S.C. 2000e(f)) and section 101(4) of the Americans with
     Disabilities Act of 1990 (42 U.S.C. 12111(4)) are each amended by adding
     at the end the following: 'With respect to employment in a foreign country,
     such term includes an individual who is a citizen of the United States.'

(b) EXEMPTION—

     (1) CIVIL RIGHTS ACT OF 1964—Section 702 of the Civil Rights Act of
          1964 (42 U.S.C. 2000e-1) is amended—

     (A) by inserting '(a)' after 'SEC. 702.'; and

     (B) by adding at the end the following:

          '(b) It shall not be unlawful under section 703 or 704 for an employer
                (or a corporation controlled by an employer), labor organiza-
                tion, employment agency, or joint labor-management commit-
                tee controlling apprenticeship or other training or retraining
                (including on-the-job training programs) to take any action oth-
                erwise prohibited by such section, with respect to an employee
                in a workplace in a foreign country if compliance with such
                section would cause such employer (or such corporation), such
                organization, such agency, or such committee to violate the law
                of the foreign country in which such workplace is located.

          '(c)(1) If an employer controls a corporation whose place of incorpo-
                ration is a foreign country, any practice prohibited by section
                703 or 704 engaged in by such corporation shall be presumed
                to be engaged in by such employer.

          '(2) Sections 703 and 704 shall not apply with respect to the for-
                eign operations of an employer that is a foreign person not con-
                trolled by an American employer.

          '(3) For purposes of this subsection, the determination of whether
                an employer controls a corporation shall be based on—

                '(A)   the interrelation of operations;

'(B)   the common management;
'(C)   the centralized control of labor relations; and
'(D)   the common ownership or financial control, of the employer and the corporation.'

(2) AMERICANS WITH DISABILITIES ACT OF 1990—Section 102 of the Americans with Disabilities Act of 1990 (42 U.S.C. 12112) is amended—

(A)   by redesignating subsection (c) as subsection (d); and
(B)   by inserting after subsection (b) the following new subsection:

'(c)   COVERED ENTITIES IN FOREIGN COUNTRIES—
'(1)   IN GENERAL—It shall not be unlawful under this section for a covered entity to take any action that constitutes discrimination under this section with respect to an employee in a workplace in a foreign country if compliance with this section would cause such covered entity to violate the law of the foreign country in which such workplace is located.
'(2)   CONTROL OF CORPORATION—
'(A)   PRESUMPTION—If an employer controls a corporation whose place of incorporation is a foreign country, any practice that constitutes discrimination under this section and is engaged in by such corporation shall be presumed to be engaged in by such employer.
'(B)   EXCEPTION—This section shall not apply with respect to the foreign operations of an employer that is a foreign person not controlled by an American employer.
'(C)   DETERMINATION—For purposes of this paragraph, the determination of whether an employer controls a corporation shall be based on—
'(i)   the interrelation of operations;
'(ii)   the common management;
'(iii)   the centralized control of labor relations; and
'(iv)   the common ownership or financial control, of the employer and the corporation.'

(c) APPLICATION OF AMENDMENTS—The amendments made by this section shall not apply with respect to conduct occurring before the date of the enactment of this Act.

Sec. 110. Technical Assistance Training Institute.

(a) TECHNICAL ASSISTANCE—Section 705 of the Civil Rights Act of 1964 (42 U.S.C. 2000e-4) is amended by adding at the end the following new subsection:

'(j)(1)   The Commission shall establish a Technical Assistance Training Institute, through which the Commission shall provide technical assistance and training regarding the laws and regulations enforced by the Commission.

    '(2) An employer or other entity covered under this title shall not be excused from compliance with the requirements of this title because of any failure to receive technical assistance under this subsection.

    '(3) There are authorized to be appropriated to carry out this subsection such sums as may be necessary for fiscal year 1992.'

  (b) EFFECTIVE DATE—The amendment made by this section shall take effect on the date of the enactment of this Act.

Sec. 111. Education and Outreach.

Section 705(h) of the Civil Rights Act of 1964 (42 U.S.C. 2000e-4(h)) is amended—

  (1) by inserting '(1)' after '(h)'; and

  (2) by adding at the end the following new paragraph:

    '(2) In exercising its powers under this title, the Commission shall carry out educational and outreach activities (including dissemination of information in languages other than English) targeted to—

      '(A) individuals who historically have been victims of employment discrimination and have not been equitably served by the Commission; and

      '(B) individuals on whose behalf the Commission has authority to enforce any other law prohibiting employment discrimination, concerning rights and obligations under this title or such law, as the case may be.'

Sec. 112. Expansion of Right to Challenge Discriminatory Seniority Systems.

Section 706(e) of the Civil Rights Act of 1964 (42 U.S.C. 2000e-5(e)) is amended—

  (1) by inserting '(1)' before 'A charge under this section'; and

  (2) by adding at the end the following new paragraph:

  'For purposes of this section, an unlawful employment practice occurs, with respect to a seniority system that has been adopted for an intentionally discriminatory purpose in violation of this title (whether or not that discriminatory purpose is apparent on the face of the seniority provision), when the seniority system is adopted, when an individual becomes subject to the seniority system, or when a person aggrieved is injured by the application of the seniority system or provision of the system.'

Sec. 113. Authorizing Award of Expert Fees.

  (a) REVISED STATUTES—Section 722 of the Revised Statutes is amended—

    (1) by designating the first and second sentences as subsections (a) and (b), respectively, and indenting accordingly; and

(2) by adding at the end the following new subsection:

    '(c) In awarding an attorney's fee under subsection (b) in any action or proceeding to enforce a provision of sections 1977 or 1977A of the Revised Statutes, the court, in its discretion, may include expert fees as part of the attorney's fee.'

(b) CIVIL RIGHTS ACT OF 1964—Section 706(k) of the Civil Rights Act of 1964 (42 U.S.C. 2000e-5(k)) is amended by inserting '(including expert fees)' after 'attorney's fee'.

Sec. 114. Providing for Interest and Extending the Statute of Limitations in Actions Against the Federal Government.

Section 717 of the Civil Rights Act of 1964 (42 U.S.C. 2000e-16) is amended—

(1) in subsection (c), by striking 'thirty days' and inserting '90 days'; and
(2) in subsection (d), by inserting before the period ', and the same interest to compensate for delay in payment shall be available as in cases involving nonpublic parties.'

Sec. 115. Notice of Limitations Period Under the Age Discrimination in Employment Act of 1967.

Section 7(e) of the Age Discrimination in Employment Act of 1967 (29 U.S.C. 626(e)) is amended—

(1) by striking paragraph (2);
(2) by striking the paragraph designation in paragraph (1);
(3) by striking 'Sections 6 and' and inserting 'Section'; and
(4) by adding at the end the following:

'If a charge filed with the Commission under this Act is dismissed or the proceedings of the Commission are otherwise terminated by the Commission, the Commission shall notify the person aggrieved. A civil action may be brought under this section by a person defined in section 11(a) against the respondent named in the charge within 90 days after the date of the receipt of such notice.'

Sec. 116. Lawful Court-Ordered Remedies, Affirmative Action, and Conciliation Agreements Not Affected.

Nothing in the amendments made by this title shall be construed to affect court-ordered remedies, affirmative action, or conciliation agreements, that are in accordance with the law.

Sec. 117. Coverage of House of Representatives and the Agencies of the Legislative Branch.

(a) COVERAGE OF THE HOUSE OF REPRESENTATIVES—

(1) IN GENERAL—Notwithstanding any provision of title VII of the Civil
Rights Act of 1964 (42 U.S.C. 2000e et seq.) or of other law, the purposes of such title shall, subject to paragraph (2), apply in their entirety
to the House of Representatives.

(2) EMPLOYMENT IN THE HOUSE—

   (A) APPLICATION—The rights and protections under title VII of the
   Civil Rights Act of 1964 (42 U.S.C. 2000e et seq.) shall, subject to
   subparagraph (B), apply with respect to any employee in an employment position in the House of Representatives and any employing
   authority of the House of Representatives.

   (B) ADMINISTRATION—

      (i)   IN GENERAL—In the administration of this paragraph, the
      remedies and procedures made applicable pursuant to the reso-
      lution described in clause (ii) shall apply exclusively.

      (ii) RESOLUTION—The resolution referred to in clause (i) is the
      Fair Employment Practices Resolution (House Resolution 558 of
      the One Hundredth Congress, as agreed to October 4, 1988), as
      incorporated into the Rules of the House of Representatives of
      the One Hundred Second Congress as Rule LI, or any other pro-
      vision that continues in effect the provisions of such resolution.

   (C) EXERCISE OF RULEMAKING POWER—The provisions of subparagraph (B) are enacted by the House of Representatives as an
   exercise of the rulemaking power of the House of Representatives,
   with full recognition of the right of the House to change its rules, in
   the same manner, and to the same extent as in the case of any other
   rule of the House.

(b) INSTRUMENTALITIES OF CONGRESS—

   (1) IN GENERAL—The rights and protections under this title and title VII
   of the Civil Rights Act of 1964 (42 U.S.C. 2000e et seq.) shall, subject
   to paragraph (2), apply with respect to the conduct of each instrumentality of the Congress.

   (2) ESTABLISHMENT OF REMEDIES AND PROCEDURES BY
   INSTRUMENTALITIES—The chief official of each instrumentality
   of the Congress shall establish remedies and procedures to be utilized
   with respect to the rights and protections provided pursuant to paragraph
   (1). Such remedies and procedures shall apply exclusively, except for the
   employees who are defined as Senate employees, in section 301(c)(1).

   (3) REPORT TO CONGRESS—The chief official of each instrumentality
   of the Congress shall, after establishing remedies and procedures for
   purposes of paragraph (2), submit to the Congress a report describing
   the remedies and procedures.

   (4) DEFINITION OF INSTRUMENTALITIES—For purposes of this
   section, instrumentalities of the Congress include the following: the
   Architect of the Capitol, the Congressional Budget Office, the General
   Accounting Office, the Government Printing Office, the Office of
   Technology Assessment, and the United States Botanic Garden.

(5) CONSTRUCTION—Nothing in this section shall alter the enforcement procedures for individuals protected under section 717 of title VII for the Civil Rights Act of 1964 (42 U.S.C. 2000e-16).

Sec. 118. Alternative Means of Dispute Resolution.

Where appropriate and to the extent authorized by law, the use of alternative means of dispute resolution, including settlement negotiations, conciliation, facilitation, mediation, factfinding, minitrials, and arbitration, is encouraged to resolve disputes arising under the Acts or provisions of Federal law amended by this title.

TITLE II—GLASS CEILING

Sec. 201. Short Title.

This title may be cited as the 'Glass Ceiling Act of 1991'.

Sec. 202. Findings and Purpose.

(a) FINDINGS—Congress finds that—
  (1) despite a dramatically growing presence in the workplace, women and minorities remain underrepresented in management and decisionmaking positions in business;
  (2) artificial barriers exist to the advancement of women and minorities in the workplace;
  (3) United States corporations are increasingly relying on women and minorities to meet employment requirements and are increasingly aware of the advantages derived from a diverse work force;
  (4) the 'Glass Ceiling Initiative' undertaken by the Department of Labor, including the release of the report entitled 'Report on the Glass Ceiling Initiative', has been instrumental in raising public awareness of—
    (A) the underrepresentation of women and minorities at the management and decisionmaking levels in the United States work force;
    (B) the underrepresentation of women and minorities in line functions in the United States work force;
    (C) the lack of access for qualified women and minorities to credential-building developmental opportunities; and
    (D) the desirability of eliminating artificial barriers to the advancement of women and minorities to such levels;
  (5) the establishment of a commission to examine issues raised by the Glass Ceiling Initiative would help—
    (A) focus greater attention on the importance of eliminating artificial barriers to the advancement of women and minorities to management and decisionmaking positions in business; and
    (B) promote work force diversity;

(6) a comprehensive study that includes analysis of the manner in which management and decisionmaking positions are filled, the developmental and skill-enhancing practices used to foster the necessary qualifications for advancement, and the compensation programs and reward structures utilized in the corporate sector would assist in the establishment of practices and policies promoting opportunities for, and eliminating artificial barriers to, the advancement of women and minorities to management and decisionmaking positions; and

(7) a national award recognizing employers whose practices and policies promote opportunities for, and eliminate artificial barriers to, the advancement of women and minorities will foster the advancement of women and minorities into higher level positions by—

    (A) helping to encourage United States companies to modify practices and policies to promote opportunities for, and eliminate artificial barriers to, the upward mobility of women and minorities; and

    (B) providing specific guidance for other United States employers that wish to learn how to revise practices and policies to improve the access and employment opportunities of women and minorities.

(b) PURPOSE—The purpose of this title is to establish—

    (1) a Glass Ceiling Commission to study—

        (A) the manner in which business fills management and decisionmaking positions;

        (B) the developmental and skill-enhancing practices used to foster the necessary qualifications for advancement into such positions; and

        (C) the compensation programs and reward structures currently utilized in the workplace; and

    (2) an annual award for excellence in promoting a more diverse skilled work force at the management and decisionmaking levels in business.

SEC. 203. ESTABLISHMENT OF GLASS CEILING COMMISSION.

(a) IN GENERAL—There is established a Glass Ceiling Commission (referred to in this title as the 'Commission'), to conduct a study and prepare recommendations concerning—

    (1) eliminating artificial barriers to the advancement of women and minorities; and

    (2) increasing the opportunities and developmental experiences of women and minorities to foster advancement of women and minorities to management and decisionmaking positions in business.

(b) MEMBERSHIP—

    (1) COMPOSITION—The Commission shall be composed of 21 members, including—

        (A) six individuals appointed by the President;

        (B) six individuals appointed jointly by the Speaker of the House of Representatives and the Majority Leader of the Senate;

    (C)   one individual appointed by the Majority Leader of the House of Representatives;

    (D)   one individual appointed by the Minority Leader of the House of Representatives;

    (E)   one individual appointed by the Majority Leader of the Senate;

    (F)   one individual appointed by the Minority Leader of the Senate;

    (G)   two Members of the House of Representatives appointed jointly by the Majority Leader and the Minority Leader of the House of Representatives;

    (H)   two Members of the Senate appointed jointly by the Majority Leader and the Minority Leader of the Senate; and

    (I)   the Secretary of Labor.

  (2) CONSIDERATIONS—In making appointments under subparagraphs (A) and (B) of paragraph (1), the appointing authority shall consider the background of the individuals, including whether the individuals—

    (A)   are members of organizations representing women and minorities, and other related interest groups;

    (B)   hold management or decisionmaking positions in corporations or other business entities recognized as leaders on issues relating to equal employment opportunity; and

    (C)   possess academic expertise or other recognized ability regarding employment issues.

  (3) BALANCE—In making the appointments under subparagraphs (A) and (B) of paragraph (1), each appointing authority shall seek to include an appropriate balance of appointees from among the groups of appointees described in subparagraphs (A), (B), and (C) of paragraph (2).

(c) CHAIRPERSON—The Secretary of Labor shall serve as the Chairperson of the Commission.

(d) TERM OF OFFICE—Members shall be appointed for the life of the Commission.

(e) VACANCIES—Any vacancy occurring in the membership of the Commission shall be filled in the same manner as the original appointment for the position being vacated. The vacancy shall not affect the power of the remaining members to execute the duties of the Commission.

(f) MEETINGS—

  (1) MEETINGS PRIOR TO COMPLETION OF REPORT—The Commission shall meet not fewer than five times in connection with and pending the completion of the report described in section 204(b). The Commission shall hold additional meetings if the Chairperson or a majority of the members of the Commission request the additional meetings in writing.

  (2) MEETINGS AFTER COMPLETION OF REPORT—The Commission shall meet once each year after the completion of the report described in section 204(b). The Commission shall hold additional meetings if the Chairperson or a majority of the members of the Commission request the additional meetings in writing.

(g) QUORUM—A majority of the Commission shall constitute a quorum for the transaction of business.

(h) COMPENSATION AND EXPENSES—

  (1) COMPENSATION—Each member of the Commission who is not an employee of the Federal Government shall receive compensation at the daily equivalent of the rate specified for level V of the Executive Schedule under section 5316 of title 5, United States Code, for each day the member is engaged in the performance of duties for the Commission, including attendance at meetings and conferences of the Commission, and travel to conduct the duties of the Commission.

  (2) TRAVEL EXPENSES—Each member of the Commission shall receive travel expenses, including per diem in lieu of subsistence, at rates authorized for employees of agencies under subchapter I of chapter 57 of title 5, United States Code, for each day the member is engaged in the performance of duties away from the home or regular place of business of the member.

  (3) EMPLOYMENT STATUS—A member of the Commission, who is not otherwise an employee of the Federal Government, shall not be deemed to be an employee of the Federal Government except for the purposes of—

    (A) the tort claims provisions of chapter 171 of title 28, United States Code; and

    (B) subchapter I of chapter 81 of title 5, United States Code, relating to compensation for work injuries.

SEC. 204. RESEARCH ON ADVANCEMENT OF WOMEN AND MINORITIES TO MANAGEMENT AND DECISIONMAKING POSITIONS IN BUSINESS.

(a) ADVANCEMENT STUDY—The Commission shall conduct a study of opportunities for, and artificial barriers to, the advancement of women and minorities to management and decisionmaking positions in business. In conducting the study, the Commission shall—

  (1) examine the preparedness of women and minorities to advance to management and decisionmaking positions in business;

  (2) examine the opportunities for women and minorities to advance to management and decisionmaking positions in business;

  (3) conduct basic research into the practices, policies, and manner in which management and decisionmaking positions in business are filled;

  (4) conduct comparative research of businesses and industries in which women and minorities are promoted to management and decisionmaking positions, and businesses and industries in which women and minorities are not promoted to management and decisionmaking positions;

  (5) compile a synthesis of available research on programs and practices that have successfully led to the advancement of women and minorities

to management and decisionmaking positions in business, including training programs, rotational assignments, developmental programs, reward programs, employee benefit structures, and family leave policies; and

(6) examine any other issues and information relating to the advancement of women and minorities to management and decisionmaking positions in business.

(b) REPORT—Not later than 15 months after the date of the enactment of this Act, the Commission shall prepare and submit to the President and the appropriate committees of Congress a written report containing—

(1) the findings and conclusions of the Commission resulting from the study conducted under subsection (a); and

(2) recommendations based on the findings and conclusions described in paragraph (1) relating to the promotion of opportunities for, and elimination of artificial barriers to, the advancement of women and minorities to management and decisionmaking positions in business, including recommendations for—

(A) policies and practices to fill vacancies at the management and decisionmaking levels;

(B) developmental practices and procedures to ensure that women and minorities have access to opportunities to gain the exposure, skills, and expertise necessary to assume management and decisionmaking positions;

(C) compensation programs and reward structures utilized to reward and retain key employees; and

(D) the use of enforcement (including such enforcement techniques as litigation, complaint investigations, compliance reviews, conciliation, administrative regulations, policy guidance, technical assistance, training, and public education) of Federal equal employment opportunity laws by Federal agencies as a means of eliminating artificial barriers to the advancement of women and minorities in employment.

(c) ADDITIONAL STUDY—The Commission may conduct such additional study of the advancement of women and minorities to management and decisionmaking positions in business as a majority of the members of the Commission determines to be necessary.

SEC. 205. ESTABLISHMENT OF THE NATIONAL AWARD FOR DIVERSITY AND EXCELLENCE IN AMERICAN EXECUTIVE MANAGEMENT.

(a) IN GENERAL—There is established the National Award for Diversity and Excellence in American Executive Management, which shall be evidenced by a medal bearing the inscription 'Frances Perkins-Elizabeth Hanford Dole National Award for Diversity and Excellence in American Executive Management'. The medal shall be of such design and materials, and bear such additional inscriptions, as the Commission may prescribe.

(b) CRITERIA FOR QUALIFICATION—To qualify to receive an award under this section a business shall—

    (1) submit a written application to the Commission, at such time, in such manner, and containing such information as the Commission may require, including at a minimum information that demonstrates that the business has made substantial effort to promote the opportunities and developmental experiences of women and minorities to foster advancement to management and decisionmaking positions within the business, including the elimination of artificial barriers to the advancement of women and minorities, and deserves special recognition as a consequence; and

    (2) meet such additional requirements and specifications as the Commission determines to be appropriate.

(c) MAKING AND PRESENTATION OF AWARD—

    (1) AWARD—After receiving recommendations from the Commission, the President or the designated representative of the President shall annually present the award described in subsection (a) to businesses that meet the qualifications described in subsection (b).

    (2) PRESENTATION—The President or the designated representative of the President shall present the award with such ceremonies as the President or the designated representative of the President may determine to be appropriate.

    (3) PUBLICITY—A business that receives an award under this section may publicize the receipt of the award and use the award in its advertising, if the business agrees to help other United States businesses improve with respect to the promotion of opportunities and developmental experiences of women and minorities to foster the advancement of women and minorities to management and decisionmaking positions.

(d) BUSINESS—For the purposes of this section, the term 'business' includes—

    (1)(A) a corporation including nonprofit corporations;

      (B) a partnership;

      (C) a professional association;

      (D) a labor organization; and

      (E) a business entity similar to an entity described in subparagraphs (A) through (D);

    (2) an education referral program, a training program, such as an apprenticeship or management training program or a similar program; and

    (3) a joint program formed by a combination of any entities described in paragraph 1 or 2.

SEC. 206. POWERS OF THE COMMISSION.

(a) IN GENERAL—The Commission is authorized to—

    (1) hold such hearings and sit and act at such times;

    (2) take such testimony;

    (3) have such printing and binding done;

    (4) enter into such contracts and other arrangements;

(5) make such expenditures; and

(6) take such other actions;

  as the Commission may determine to be necessary to carry out the duties of the Commission.

(b) OATHS—Any member of the Commission may administer oaths or affirmations to witnesses appearing before the Commission.

(c) OBTAINING INFORMATION FROM FEDERAL AGENCIES—The Commission may secure directly from any Federal agency such information as the Commission may require to carry out its duties.

(d) VOLUNTARY SERVICE—Notwithstanding section 1342 of title 31, United States Code, the Chairperson of the Commission may accept for the Commission voluntary services provided by a member of the Commission.

(e) GIFTS AND DONATIONS—The Commission may accept, use, and dispose of gifts or donations of property in order to carry out the duties of the Commission.

(f) USE OF MAIL—The Commission may use the United States mails in the same manner and under the same conditions as Federal agencies.

SEC. 207. CONFIDENTIALITY OF INFORMATION.

(a) Individual Business Information—

(1) IN GENERAL—Except as provided in paragraph (2), and notwithstanding section 552 of title 5, United States Code, in carrying out the duties of the Commission, including the duties described in sections 204 and 205, the Commission shall maintain the confidentiality of all information that concerns—

(A) the employment practices and procedures of individual businesses; or

(B) individual employees of the businesses.

(2) CONSENT—The content of any information described in paragraph (1) may be disclosed with the prior written consent of the business or employee, as the case may be, with respect to which the information is maintained.

(b) AGGREGATE INFORMATION—In carrying out the duties of the Commission, the Commission may disclose—

(1) information about the aggregate employment practices or procedures of a class or group of businesses; and

(2) information about the aggregate characteristics of employees of the businesses, and related aggregate information about the employees.

SEC. 208. STAFF AND CONSULTANTS.

(a) STAFF—

(1) APPOINTMENT AND COMPENSATION—The Commission may appoint and determine the compensation of such staff as the Commission determines to be necessary to carry out the duties of the Commission.

(2) LIMITATIONS—The rate of compensation for each staff member shall not exceed the daily equivalent of the rate specified for level V of the Executive Schedule under section 5316 of title 5, United States Code for each day the staff member is engaged in the performance of duties for the Commission. The Commission may otherwise appoint and determine the compensation of staff without regard to the provisions of title 5, United States Code, that govern appointments in the competitive service, and the provisions of chapter 51 and subchapter III of chapter 53 of title 5, United States Code, that relate to classification and General Schedule pay rates.

(b) EXPERTS AND CONSULTANTS—The Chairperson of the Commission may obtain such temporary and intermittent services of experts and consultants and compensate the experts and consultants in accordance with section 3109(b) of title 5, United States Code, as the Commission determines to be necessary to carry out the duties of the Commission.

(c) DETAIL OF FEDERAL EMPLOYEES—On the request of the Chairperson of the Commission, the head of any Federal agency shall detail, without reimbursement, any of the personnel of the agency to the Commission to assist the Commission in carrying out its duties. Any detail shall not interrupt or otherwise affect the civil service status or privileges of the Federal employee.

(d) TECHNICAL ASSISTANCE—On the request of the Chairperson of the Commission, the head of a Federal agency shall provide such technical assistance to the Commission as the Commission determines to be necessary to carry out its duties.

SEC. 209. AUTHORIZATION OF APPROPRIATIONS.

There are authorized to be appropriated to the Commission such sums as may be necessary to carry out the provisions of this title. The sums shall remain available until expended, without fiscal year limitation.

SEC. 210. TERMINATION.

(a) COMMISSION—Notwithstanding section 15 of the Federal Advisory Committee Act (5 U.S.C. App.), the Commission shall terminate 4 years after the date of the enactment of this Act.

(b) AWARD—The authority to make awards under section 205 shall terminate 4 years after the date of the enactment of this Act.

TITLE III—GOVERNMENT EMPLOYEE RIGHTS

SEC. 301. GOVERNMENT EMPLOYEE RIGHTS ACT OF 1991.

(a) SHORT TITLE—This title may be cited as the 'Government Employee Rights Act of 1991'.

(b) PURPOSE—The purpose of this title is to provide procedures to protect the right of Senate and other government employees, with respect to their public employment, to be free of discrimination on the basis of race, color, religion, sex, national origin, age, or disability.

(c) DEFINITIONS—For purposes of this title:

   (1) SENATE EMPLOYEE—The term 'Senate employee' or 'employee' means—

      (A) any employee whose pay is disbursed by the Secretary of the Senate;

      (B) any employee of the Architect of the Capitol who is assigned to the Senate Restaurants or to the Superintendent of the Senate Office Buildings;

      (C) any applicant for a position that will last 90 days or more and that is to be occupied by an individual described in subparagraph (A) or (B); or

      (D) any individual who was formerly an employee described in subparagraph (A) or (B) and whose claim of a violation arises out of the individual's Senate employment.

   (2) HEAD OF EMPLOYING OFFICE—The term 'head of employing office' means the individual who has final authority to appoint, hire, discharge, and set the terms, conditions or privileges of the Senate employment of an employee.

   (3) VIOLATION—The term 'violation' means a practice that violates section 302 of this title.

Sec. 302. Discriminatory Practices Prohibited.

All personnel actions affecting employees of the Senate shall be made free from any discrimination based on—

(1) race, color, religion, sex, or national origin, within the meaning of section 717 of the Civil Rights Act of 1964 (42 U.S.C. 2000e-16);

(2) age, within the meaning of section 15 of the Age Discrimination in Employment Act of 1967 (29 U.S.C. 633a); or

(3) handicap or disability, within the meaning of section 501 of the Rehabilitation Act of 1973 (29 U.S.C. 791) and sections 102–104 of the Americans with Disabilities Act of 1990 (42 U.S.C. 12112-14).

Sec. 303. Establishment of Office of Senate Fair Employment Practices.

(a) IN GENERAL—There is established, as an office of the Senate, the Office of Senate Fair Employment Practices (referred to in this title as the 'Office'), which shall—

   (1) administer the processes set forth in sections 305 through 307;

   (2) implement programs for the Senate to heighten awareness of employee rights in order to prevent violations from occurring.

(b) DIRECTOR—

(1) IN GENERAL—The Office shall be headed by a Director (referred to in this title as the 'Director') who shall be appointed by the President pro tempore, upon the recommendation of the Majority Leader in consultation with the Minority Leader. The appointment shall be made without regard to political affiliation and solely on the basis of fitness to perform the duties of the position. The Director shall be appointed for a term of service which shall expire at the end of the Congress following the Congress during which the Director is appointed. A Director may be reappointed at the termination of any term of service. The President pro tempore, upon the joint recommendation of the Majority Leader in consultation with the Minority Leader, may remove the Director at any time.

(2) SALARY—The President pro tempore, upon the recommendation of the Majority Leader in consultation with the Minority Leader, shall establish the rate of pay for the Director. The salary of the Director may not be reduced during the employment of the Director and shall be increased at the same time and in the same manner as fixed statutory salary rates within the Senate are adjusted as a result of annual comparability increases.

(3) ANNUAL BUDGET—The Director shall submit an annual budget request for the Office to the Committee on Appropriations.

(4) APPOINTMENT OF DIRECTOR—The first Director shall be appointed and begin service within 90 days after the date of enactment of this Act, and thereafter the Director shall be appointed and begin service within 30 days after the beginning of the session of the Congress immediately following the termination of a Director's term of service or within 60 days after a vacancy occurs in the position.

(c) STAFF OF THE OFFICE—

(1) APPOINTMENT—The Director may appoint and fix the compensation of such additional staff, including hearing officers, as are necessary to carry out the purposes of this title.

(2) DETAILEES—The Director may, with the prior consent of the Government department or agency concerned and the Committee on Rules and Administration, use on a reimbursable or nonreimbursable basis the services of any such department or agency, including the services of members or personnel of the General Accounting Office Personnel Appeals Board.

(3) CONSULTANTS—In carrying out the functions of the Office, the Director may procure the temporary (not to exceed 1 year) or intermittent services of individual consultants, or organizations thereof, in the same manner and under the same conditions as a standing committee of the Senate may procure such services under section 202(i) of the Legislative Reorganization Act of 1946 (2 U.S.C. 72a(i)).

(d) EXPENSES OF THE OFFICE—In fiscal year 1992, the expenses of the Office shall be paid out of the Contingent Fund of the Senate from the appropriation account Miscellaneous Items. Beginning in fiscal year 1993, and for each fiscal year thereafter, there is authorized to be appropriated for the expenses of the Office such sums as shall be necessary to carry out its functions. In all cases, expenses shall be paid out of the Contingent Fund of the Senate upon vouchers approved by the Director, except that a voucher shall not be required for—

(1) the disbursement of salaries of employees who are paid at an annual rate;
(2) the payment of expenses for telecommunications services provided by the Telecommunications Department, Sergeant at Arms, United States Senate;
(3) the payment of expenses for stationery supplies purchased through the Keeper of the Stationery, United States Senate;
(4) the payment of expenses for postage to the Postmaster, United States Senate; and
(5) the payment of metered charges on copying equipment provided by the Sergeant at Arms, United States Senate.

The Secretary of the Senate is authorized to advance such sums as may be necessary to defray the expenses incurred in carrying out this title. Expenses of the Office shall include authorized travel for personnel of the Office.

(e) RULES OF THE OFFICE—The Director shall adopt rules governing the procedures of the Office, including the procedures of hearing boards, which rules shall be submitted to the President pro tempore for publication in the Congressional Record. The rules may be amended in the same manner. The Director may consult with the Chairman of the Administrative Conference of the United States on the adoption of rules.

(f) REPRESENTATION BY THE SENATE LEGAL COUNSEL—For the purpose of representation by the Senate Legal Counsel, the Office shall be deemed a committee, within the meaning of title VII of the Ethics in Government Act of 1978 (2 U.S.C. 288, et seq.).

Sec. 304. Senate Procedure for Consideration of Alleged Violations.

The Senate procedure for consideration of alleged violations consists of 4 steps as follows:

(1) Step I, counseling, as set forth in section 305.
(2) Step II, mediation, as set forth in section 306.
(3) Step III, formal complaint and hearing by a hearing board, as set forth in section 307.
(4) Step IV, review of a hearing board decision, as set forth in section 308 or 309.

Sec. 305. Step I: Counseling.

(a) IN GENERAL—A Senate employee alleging a violation may request counseling by the Office. The Office shall provide the employee with all relevant information with respect to the rights of the employee. A request for counseling shall be made not later than 180 days after the alleged violation forming the basis of the request for counseling occurred. No request for counseling may be made until 10 days after the first Director begins service pursuant to section 303(b)(4).

(b) PERIOD OF COUNSELING—The period for counseling shall be 30 days unless the employee and the Office agree to reduce the period. The period shall begin on the date the request for counseling is received.

(c) EMPLOYEES OF THE ARCHITECT OF THE CAPITOL AND CAPITOL POLICE—In the case of an employee of the Architect of the Capitol or an employee who is a member of the Capitol Police, the Director may refer the employee to the Architect of the Capitol or the Capitol Police Board for resolution of the employee's complaint through the internal grievance procedures of the Architect of the Capitol or the Capitol Police Board for a specific period of time, which shall not count against the time available for counseling or mediation under this title.

Sec. 306. Step II: Mediation.

(a) IN GENERAL—Not later than 15 days after the end of the counseling period, the employee may file a request for mediation with the Office. Mediation may include the Office, the employee, and the employing office in a process involving meetings with the parties separately or jointly for the purpose of resolving the dispute between the employee and the employing office.

(b) MEDIATION PERIOD—The mediation period shall be 30 days beginning on the date the request for mediation is received and may be extended for an additional 30 days at the discretion of the Office. The Office shall notify the employee and the head of the employing office when the mediation period has ended.

Sec. 307. Step III: Formal Complaint and Hearing.

(a) FORMAL COMPLAINT AND REQUEST FOR HEARING—Not later than 30 days after receipt by the employee of notice from the Office of the end of the mediation period, the Senate employee may file a formal complaint with the Office. No complaint may be filed unless the employee has made a timely request for counseling and has completed the procedures set forth in sections 305 and 306.

(b) HEARING BOARD—A board of 3 independent hearing officers (referred to in this title as 'hearing board'), who are not Senators or officers or employees of the Senate, chosen by the Director (one of whom shall be designated by the Director as the presiding hearing officer) shall be assigned to consider each complaint filed under this section. The Director shall appoint hearing officers after considering any candidates who are recommended to the Director by the Federal Mediation and Conciliation Service, the Administrative Conference of the United States, or organizations composed primarily of individuals experienced in adjudicating or arbitrating personnel matters. A hearing board shall act by majority vote.

(c) DISMISSAL OF FRIVOLOUS CLAIMS—Prior to a hearing under subsection (d), a hearing board may dismiss any claim that it finds to be frivolous.

(d) HEARING—A hearing shall be conducted—
   (1) in closed session on the record by a hearing board;
   (2) no later than 30 days after filing of the complaint under subsection (a), except that the Office may, for good cause, extend up to an additional 60 days the time for conducting a hearing; and
   (3) except as specifically provided in this title and to the greatest extent practicable, in accordance with the principles and procedures set forth in sections 554 through 557 of title 5, United States Code.

(e) DISCOVERY—Reasonable prehearing discovery may be permitted at the discretion of the hearing board.

(f) SUBPOENA—
   (1) AUTHORIZATION—A hearing board may authorize subpoenas, which shall be issued by the presiding hearing officer on behalf of the hearing board, for the attendance of witnesses at proceedings of the hearing board and for the production of correspondence, books, papers, documents, and other records.
   (2) OBJECTIONS—If a witness refuses, on the basis of relevance, privilege, or other objection, to testify in response to a question or to produce records in connection with the proceedings of a hearing board, the hearing board shall rule on the objection. At the request of the witness, the employee, or employing office, or on its own initiative, the hearing board may refer the objection to the Select Committee on Ethics for a ruling.
   (3) ENFORCEMENT—The Select Committee on Ethics may make to the Senate any recommendations by report or resolution, including recommendations for criminal or civil enforcement by or on behalf of the Office, which the Select Committee on Ethics may consider appropriate with respect to—
      (A) the failure or refusal of any person to appear in proceedings under this or to produce records in obedience to a subpoena or order of the hearing board; or
      (B) the failure or refusal of any person to answer questions during his or her appearance as a witness in a proceeding under this section.
      For purposes of section 1365 of title 28, United States Code, the Office shall be deemed to be a committee of the Senate.

(g) DECISION—The hearing board shall issue a written decision as expeditiously as possible, but in no case more than 45 days after the conclusion of the hearing. The written decision shall be transmitted by the Office to the employee and the employing office. The decision shall state the issues raised by the complaint, describe the evidence in the record, and contain a determination as to whether a violation has occurred.

(h) REMEDIES—If the hearing board determines that a violation has occurred, it shall order such remedies as would be appropriate if awarded under section 706 (g) and (k) of the Civil Rights Act of 1964 (42 U.S.C. 2000e-5 (g) and (k)), and may also order the award of such compensatory damages as would be appropriate if awarded under section 1977 and section 1977A (a) and (b)(2) of the Revised Statutes (42 U.S.C. 1981 and 1981A (a) and (b)(2)). In the case of a determination that a violation based on age has occurred, the hearing board shall order such remedies as would be appropriate if awarded under section 15(c) of the Age Discrimination in Employment Act of 1967 (29 U.S.C. 633a(c)). Any order requiring the payment of money must be approved by a Senate resolution reported by the Committee on Rules and Administration. The hearing board shall have no authority to award punitive damages.

(i) PRECEDENT AND INTERPRETATIONS—Hearing boards shall be guided by judicial decisions under statutes referred to in section 302 and subsection (h) of this section, as well as the precedents developed by the Select Committee on Ethics under section 308, and other Senate precedents.

SEC. 308. REVIEW BY THE SELECT COMMITTEE ON ETHICS.

(a) IN GENERAL—An employee or the head of an employing office may request that the Select Committee on Ethics (referred to in this section as the 'Committee'), or such other entity as the Senate may designate, review a decision under section 307, including any decision following a remand under subsection (c), by filing a request for review with the Office not later than 10 days after the receipt of the decision of a hearing board. The Office, at the discretion of the Director, on its own initiative and for good cause, may file a request for review by the Committee of a decision of a hearing board not later than 5 days after the time for the employee or employing office to file a request for review has expired. The Office shall transmit a copy of any request for review to the Committee and notify the interested parties of the filing of the request for review.

(b) REVIEW—Review under this section shall be based on the record of the hearing board. The Committee shall adopt and publish in the Congressional Record procedures for requests for review under this section.

(c) REMAND—Within the time for a decision under subsection (d), the Committee may remand a decision no more than one time to the hearing board for the purpose of supplementing the record or for further consideration.

(d) FINAL DECISION—

   (1) HEARING BOARD—If no timely request for review is filed under subsection (a), the Office shall enter as a final decision, the decision of the hearing board.

   (2) SELECT COMMITTEE ON ETHICS—

      (A) If the Committee does not remand under subsection (c), it shall transmit a written final decision to the Office for entry in the records of the Office. The Committee shall transmit the decision not later than

60 calendar days during which the Senate is in session after the filing of a request for review under subsection (a). The Committee may extend for 15 calendar days during which the Senate is in session the period for transmission to the Office of a final decision.

(B) The decision of the hearing board shall be deemed to be a final decision, and entered in the records of the Office as a final decision, unless a majority of the Committee votes to reverse or remand the decision of the hearing board within the time for transmission to the Office of a final decision.

(C) The decision of the hearing board shall be deemed to be a final decision, and entered in the records of the Office as a final decision, if the Committee, in its discretion, decides not to review, pursuant to a request for review under subsection (a), a decision of the hearing board, and notifies the interested parties of such decision.

(3) ENTRY OF A FINAL DECISION—The entry of a final decision in the records of the Office shall constitute a final decision for purposes of judicial review under section 309.

(e) STATEMENT OF REASONS—Any decision of the Committee under subsection (c) or subsection (d)(2)(A) shall contain a written statement of the reasons for the Committee's decision.

SEC. 309. JUDICIAL REVIEW.

(a) IN GENERAL—Any Senate employee aggrieved by a final decision under section 308(d), or any Member of the Senate who would be required to reimburse the appropriate Federal account pursuant to the section entitled 'Payments by the President or a Member of the Senate' and a final decision entered pursuant to section 308(d)(2)(B), may petition for review by the United States Court of Appeals for the Federal Circuit.

(b) LAW APPLICABLE—Chapter 158 of title 28, United States Code, shall apply to a review under this section except that—

(1) with respect to section 2344 of title 28, United States Code, service of the petition shall be on the Senate Legal Counsel rather than on the Attorney General;

(2) the provisions of section 2348 of title 28, United States Code, on the authority of the Attorney General, shall not apply;

(3) the petition for review shall be filed not later than 90 days after the entry in the Office of a final decision under section 308(d);

(4) the Office shall be an 'agency' as that term is used in chapter 158 of title 28, United States Code; and

(5) the Office shall be the respondent in any proceeding under this section.

(c) STANDARD OF REVIEW—To the extent necessary to decision and when presented, the court shall decide all relevant questions of law and interpret constitutional and statutory provisions. The court shall set aside a final decision if it is determined that the decision was—

   (1) arbitrary, capricious, an abuse of discretion, or otherwise not consistent with law;

   (2) not made consistent with required procedures; or

   (3) unsupported by substantial evidence.

     In making the foregoing determinations, the court shall review the whole record, or those parts of it cited by a party, and due account shall be taken of the rule of prejudicial error. The record on review shall include the record before the hearing board, the decision of the hearing board, and the decision, if any, of the Select Committee on Ethics.

 (d) ATTORNEY'S FEES—If an employee is the prevailing party in a proceeding under this section, attorney's fees may be allowed by the court in accordance with the standards prescribed under section 706(k) of the Civil Rights Act of 1964 (42 U.S.C. 2000e-5(k)).

Sec. 310. Resolution of Complaint.

If, after a formal complaint is filed under section 307, the employee and the head of the employing office resolve the issues involved, the employee may dismiss the complaint or the parties may enter into a written agreement, subject to the approval of the Director.

Sec. 311. Costs of Attending Hearings.

Subject to the approval of the Director, an employee with respect to whom a hearing is held under this title may be reimbursed for actual and reasonable costs of attending proceedings under sections 307 and 308, consistent with Senate travel regulations. Senate Resolution 259, agreed to August 5, 1987 (100th Congress, 1st Session), shall apply to witnesses appearing in proceedings before a hearing board.

Sec. 312. Prohibition of Intimidation.

Any intimidation of, or reprisal against, any employee by any Member, officer, or employee of the Senate, or by the Architect of the Capitol, or anyone employed by the Architect of the Capitol, as the case may be, because of the exercise of a right under this title constitutes an unlawful employment practice, which may be remedied in the same manner under this title as is a violation.

Sec. 313. Confidentiality.

 (a) COUNSELING—All counseling shall be strictly confidential except that the Office and the employee may agree to notify the head of the employing office of the allegations.

 (b) MEDIATION—All mediation shall be strictly confidential.

 (c) HEARINGS—Except as provided in subsection (d), the hearings, deliberations, and decisions of the hearing board and the Select Committee on Ethics shall be confidential.

(d) FINAL DECISION OF SELECT COMMITTEE ON ETHICS—The final decision of the Select Committee on Ethics under section 308 shall be made public if the decision is in favor of the complaining Senate employee or if the decision reverses a decision of the hearing board which had been in favor of the employee. The Select Committee on Ethics may decide to release any other decision at its discretion. In the absence of a proceeding under section 308, a decision of the hearing board that is favorable to the employee shall be made public.

(e) RELEASE OF RECORDS FOR JUDICIAL REVIEW—The records and decisions of hearing boards, and the decisions of the Select Committee on Ethics, may be made public if required for the purpose of judicial review under section 309.

SEC. 314. EXERCISE OF RULEMAKING POWER.

The provisions of this title, except for sections 309, 320, 321, and 322, are enacted by the Senate as an exercise of the rulemaking power of the Senate, with full recognition of the right of the Senate to change its rules, in the same manner, and to the same extent, as in the case of any other rule of the Senate. Notwithstanding any other provision of law, except as provided in section 309, enforcement and adjudication with respect to the discriminatory practices prohibited by section 302, and arising out of Senate employment, shall be within the exclusive jurisdiction of the United States Senate.

SEC. 315. TECHNICAL AND CONFORMING AMENDMENTS.

Section 509 of the Americans with Disabilities Act of 1990 (42 U.S.C. 12209) is amended—

(1) in subsection (a)—
  (A) by striking paragraphs (2) through (5);
  (B) by redesignating paragraphs (6) and (7) as paragraphs (2) and (3), respectively; and
  (C) in paragraph (3), as redesignated by subparagraph (B) of this paragraph—
    (i) by striking '(2) and (6)(A)' and inserting '(2)(A)', as redesignated by subparagraph (B) of this paragraph; and
    (ii) by striking '(3), (4), (5), (6)(B), and (6)(C)' and inserting '(2)'; and
(2) in subsection (c)(2), by inserting ', except for the employees who are defined as Senate employees, in section 301(c)(1) of the Civil Rights Act of 1991' after 'shall apply exclusively'.

SEC. 316. POLITICAL AFFILIATION AND PLACE OF RESIDENCE.

(a) IN GENERAL—It shall not be a violation with respect to an employee described in subsection (b) to consider the—
  (1) party affiliation;

(2) domicile; or
(3) political compatibility with the employing office, of such an employee with respect to employment decisions.
(b) DEFINITION—For purposes of this section, the term 'employee' means—
  (1) an employee on the staff of the Senate leadership;
  (2) an employee on the staff of a committee or subcommittee;
  (3) an employee on the staff of a Member of the Senate;
  (4) an officer or employee of the Senate elected by the Senate or appointed by a Member, other than those described in paragraphs (1) through (3); or
  (5) an applicant for a position that is to be occupied by an individual described in paragraphs (1) through (4).

Sec. 317. Other Review.

No Senate employee may commence a judicial proceeding to redress discriminatory practices prohibited under section 302 of this title, except as provided in this title.

Sec. 318. Other Instrumentalities of the Congress.

It is the sense of the Senate that legislation should be enacted to provide the same or comparable rights and remedies as are provided under this title to employees of instrumentalities of the Congress not provided with such rights and remedies.

Sec. 319. Rule XLII of the Standing Rules of the Senate.

(a) REAFFIRMATION—The Senate reaffirms its commitment to Rule XLII of the Standing Rules of the Senate, which provides as follows:
    'No Member, officer, or employee of the Senate shall, with respect to employment by the Senate or any office thereof—
  '(a) fail or refuse to hire an individual;
  '(b) discharge an individual; or
  '(c) otherwise discriminate against an individual with respect to promotion, compensation, or terms, conditions, or privileges of employment on the basis of such individual's race, color, religion, sex, national origin, age, or state of physical handicap.'
(b) AUTHORITY TO DISCIPLINE—Notwithstanding any provision of this title, including any provision authorizing orders for remedies to Senate employees to redress employment discrimination, the Select Committee on Ethics shall retain full power, in accordance with its authority under Senate Resolution 338, 88th Congress, as amended, with respect to disciplinary action against a Member, officer, or employee of the Senate for a violation of Rule XLII.

Sec. 320. Coverage of Presidential Appointees.

(a) IN GENERAL—

(1) APPLICATION—The rights, protections, and remedies provided pursuant to section 302 and 307(h) of this title shall apply with respect to employment of Presidential appointees.

(2) ENFORCEMENT BY ADMINISTRATIVE ACTION—Any Presidential appointee may file a complaint alleging a violation, not later than 180 days after the occurrence of the alleged violation, with the Equal Employment Opportunity Commission, or such other entity as is designated by the President by Executive Order, which, in accordance with the principles and procedures set forth in sections 554 through 557 of title 5, United States Code, shall determine whether a violation has occurred and shall set forth its determination in a final order. If the Equal Employment Opportunity Commission, or such other entity as is designated by the President pursuant to this section, determines that a violation has occurred, the final order shall also provide for appropriate relief.

(3) JUDICIAL REVIEW—

(A) IN GENERAL—Any party aggrieved by a final order under paragraph (2) may petition for review by the United States Court of Appeals for the Federal Circuit.

(B) LAW APPLICABLE—Chapter 158 of title 28, United States Code, shall apply to a review under this section except that the Equal Employment Opportunity Commission or such other entity as the President may designate under paragraph (2) shall be an 'agency' as that term is used in chapter 158 of title 28, United States Code.

(C) STANDARD OF REVIEW—To the extent necessary to decision and when presented, the reviewing court shall decide all relevant questions of law and interpret constitutional and statutory provisions. The court shall set aside a final order under paragraph (2) if it is determined that the order was—

(i) arbitrary, capricious, an abuse of discretion, or otherwise not consistent with law;

(ii) not made consistent with required procedures; or

(iii) unsupported by substantial evidence.

In making the foregoing determinations, the court shall review the whole record or those parts of it cited by a party, and due account shall be taken of the rule of prejudicial error.

(D) ATTORNEY'S FEES—If the presidential appointee is the prevailing party in a proceeding under this section, attorney's fees may be allowed by the court in accordance with the standards prescribed under section 706(k) of the Civil Rights Act of 1964 (42 U.S.C. 2000e-5(k)).

(b) PRESIDENTIAL APPOINTEE—For purposes of this section, the term 'Presidential appointee' means any officer or employee, or an applicant seeking to become an officer or employee, in any unit of the Executive Branch, including the Executive Office of the President, whether appointed by the President or by any other appointing authority in the Executive Branch, who is not already entitled to bring an action under any of the statutes referred to in section 302 but does not include any individual—

(1) whose appointment is made by and with the advice and consent of the Senate;

(2) who is appointed to an advisory committee, as defined in section 3(2) of the Federal Advisory Committee Act (5 U.S.C. App.); or

(3) who is a member of the uniformed services.

Sec. 321. Coverage of Previously Exempt State Employees.

(a) APPLICATION—The rights, protections, and remedies provided pursuant to section 302 and 307(h) of this title shall apply with respect to employment of any individual chosen or appointed, by a person elected to public office in any State or political subdivision of any State by the qualified voters thereof—

(1) to be a member of the elected official's personal staff;

(2) to serve the elected official on the policymaking level; or

(3) to serve the elected official as an immediate advisor with respect to the exercise of the constitutional or legal powers of the office.

(b) Enforcement by Administrative Action—

(1) IN GENERAL—Any individual referred to in subsection (a) may file a complaint alleging a violation, not later than 180 days after the occurrence of the alleged violation, with the Equal Employment Opportunity Commission, which, in accordance with the principles and procedures set forth in sections 554 through 557 of title 5, United States Code, shall determine whether a violation has occurred and shall set forth its determination in a final order. If the Equal Employment Opportunity Commission determines that a violation has occurred, the final order shall also provide for appropriate relief.

(2) REFERRAL TO STATE AND LOCAL AUTHORITIES—

(A) APPLICATION—Section 706(d) of the Civil Rights Act of 1964 (42 U.S.C. 2000e-5(d)) shall apply with respect to any proceeding under this section.

(B) DEFINITION—For purposes of the application described in subparagraph (A), the term 'any charge filed by a member of the Commission alleging an unlawful employment practice' means a complaint filed under this section.

(c) JUDICIAL REVIEW—Any party aggrieved by a final order under subsection (b) may obtain a review of such order under chapter 158 of title 28, United States Code. For the purpose of this review, the Equal Employment Opportunity Commission shall be an 'agency' as that term is used in chapter 158 of title 28, United States Code.

(d) STANDARD OF REVIEW—To the extent necessary to decision and when presented, the reviewing court shall decide all relevant questions of law and interpret constitutional and statutory provisions. The court shall set aside a final order under subsection (b) if it is determined that the order was—

(1) arbitrary, capricious, an abuse of discretion, or otherwise not consistent with law;

(2) not made consistent with required procedures; or

(3) unsupported by substantial evidence.

In making the foregoing determinations, the court shall review the whole record or those parts of it cited by a party, and due account shall be taken of the rule of prejudicial error.

(e) ATTORNEY'S FEES—If the individual referred to in subsection (a) is the prevailing party in a proceeding under this subsection, attorney's fees may be allowed by the court in accordance with the standards prescribed under section 706(k) of the Civil Rights Act of 1964 (42 U.S.C. 2000e-5(k)).

SEC. 322. SEVERABILITY.

Notwithstanding section 401 of this Act, if any provision of section 309 or 320(a)(3) is invalidated, both sections 309 and 320(a)(3) shall have no force and effect.

SEC. 323. PAYMENTS BY THE PRESIDENT OR A MEMBER OF THE SENATE.

The President or a Member of the Senate shall reimburse the appropriate Federal account for any payment made on his or her behalf out of such account for a violation committed under the provisions of this title by the President or Member of the Senate not later than 60 days after the payment is made.

SEC. 324. REPORTS OF SENATE COMMITTEES.

(a) Each report accompanying a bill or joint resolution of a public character reported by any committee of the Senate (except the Committee on Appropriations and the Committee on the Budget) shall contain a listing of the provisions of the bill or joint resolution that apply to Congress and an evaluation of the impact of such provisions on Congress.

(b) The provisions of this section are enacted by the Senate as an exercise of the rulemaking power of the Senate, with full recognition of the right of the Senate to change its rules, in the same manner, and to the same extent, as in the case of any other rule of the Senate.

SEC. 325. INTERVENTION AND EXPEDITED REVIEW OF CERTAIN APPEALS.

(a) INTERVENTION—Because of the constitutional issues that may be raised by section 309 and section 320, any Member of the Senate may intervene as a matter of right in any proceeding under section 309 for the sole purpose of determining the constitutionality of such section.

(b) THRESHOLD MATTER—In any proceeding under section 309 or section 320, the United States Court of Appeals for the Federal Circuit shall determine any issue presented concerning the constitutionality of such section as a threshold matter.

(c) APPEAL—

(1) IN GENERAL—An appeal may by taken directly to the Supreme Court of the United States from any interlocutory or final judgment, decree, or order issued by the United States Court of Appeals for the Federal Circuit ruling upon the constitutionality of section 309 or 320.

(2) JURISDICTION—The Supreme Court shall, if it has not previously ruled on the question, accept jurisdiction over the appeal referred to in paragraph (1), advance the appeal on the docket and expedite the appeal to the greatest extent possible.

TITLE IV—GENERAL PROVISIONS

SEC. 401. SEVERABILITY.

If any provision of this Act, or an amendment made by this Act, or the application of such provision to any person or circumstances is held to be invalid, the remainder of this Act and the amendments made by this Act, and the application of such provision to other persons and circumstances, shall not be affected.

SEC. 402. EFFECTIVE DATE.

(a) IN GENERAL—Except as otherwise specifically provided, this Act and the amendments made by this Act shall take effect upon enactment.

(b) CERTAIN DISPARATE IMPACT CASES—Notwithstanding any other provision of this Act, nothing in this Act shall apply to any disparate impact case for which a complaint was filed before March 1, 1975, and for which an initial decision was rendered after October 30, 1983.

TITLE V—CIVIL WAR SITES ADVISORY COMMISSION

SEC. 501. CIVIL WAR SITES ADVISORY COMMISSION.

Section 1205 of Public Law 101-628 is amended in subsection (a) by—

(1) striking 'Three' in paragraph (4) and inserting 'Four' in lieu thereof; and

(2) striking 'Three' in paragraph (5) and inserting 'Four' in lieu thereof.

Appendix 3: Wage and Hour Division (WHD): Minimum Wage Laws in the States—July 1, 2010

HISTORICAL TABLE

Note: Where federal and state law have different minimum wage rates, the higher standard applies.

States with minimum wage rates higher than the Federal

States with minimum wage rates the same as the Federal

American Samoa has special minimum wage rates

States with no minimum wage law

States with minimum wage rates lower than the Federal

Minimum Wage and Overtime Premium Pay Standards Applicable to Nonsupervisory NONFARM *Private Sector* Employment Under State and Federal Laws July 1, 2010.[1]

CONSOLIDATED STATE MINIMUM WAGE UPDATE TABLE

Alabama Minimum Wage Rates

ALABAMA	Future Effective Date	Basic Minimum Rate (per hour)	Premium Pay after Designated Hours[2]	
			Daily	Weekly
No state minimum wage law.				

Alaska Minimum Wage Rates

ALASKA	Future Effective Date	Basic Minimum Rate (per hour)	Premium Pay after Designated Hours[2]	
			Daily	Weekly
			8	40
		$7.75		

Under a voluntary flexible work hour plan approved by the Alaska Department of Labor, a 10 hour day, 40 hour workweek may be instituted with premium pay after 10 hours a day instead of after 8 hours.

The premium overtime pay requirement on either a daily or weekly basis is not applicable to employers of fewer than 4 employees.

AMERICAN SAMOA

American Samoa has special minimum wage rates.

Arizona Minimum Wage Rates

ARIZONA	Future Effective Date	Basic Minimum Rate (per hour)	Premium Pay after Designated Hours[2]	
			Daily	Weekly
		$7.25		

Rate is increased annually based upon a cost of living formula.

Arkansas Minimum Wage Rates

ARKANSAS	Future Effective Date	Basic Minimum Rate (per hour)	Premium Pay after Designated Hours[2]	
			Daily	Weekly
(Applicable to employers of 4 or more employees)		$6.25	N/A	40

California Minimum Wage Rates

CALIFORNIA	Future Effective Date	Basic Minimum Rate (per hour)	Premium Pay after Designated Hours[2]	
			Daily	Weekly
		$8.00	8 Over 12 (double time)	40; on 7th day: First 8 hours (time and half) Over 8 hours on 7th day (double time)

Any work in excess of eight hours in one workday and any work in excess of 40 hours in one workweek and the first eight hours worked on the seventh day of work in any one workweek shall be at the rate of one and one-half times the regular rate of pay. Any work in excess of 12 hours in one day and any work in excess of eight hours on any seventh day of a workweek shall be paid no less than twice the regular rate of pay. California Labor Code section 310. Exceptions apply to an employee working pursuant to an alternative workweek adopted pursuant to applicable Labor Code sections and for time spent commuting. (See Labor Code sections 510 for exceptions).

Colorado Minimum Wage Rates

COLORADO	Future Effective Date	Basic Minimum Rate (per hour)	Premium Pay after Designated Hours[2]	
			Daily	Weekly
		$7.24	12	40

Minimum wage rate and overtime provisions applicable to retail and service, commercial support service, food and beverage, and health and medical industries.

Rate is increased or decreased annually based upon a cost of living formula.

Connecticut Minimum Wage Rates

CONNECTICUT	Future Effective Date	Basic Minimum Rate (per hour)	Premium Pay after Designated Hours[2]	
			Daily	Weekly
				40
		$8.25		

In restaurants and hotel restaurants, for the 7th consecutive day of work, premium pay is required at time and one half the minimum rate.

The Connecticut minimum wage rate automatically increases to 1/2 of 1 percent above the rate set in the Fair Labor Standards Act if the Federal minimum wage rate equals or becomes higher than the State minimum.

Delaware Minimum Wage Rates

DELAWARE	Future Effective Date	Basic Minimum Rate (per hour)	Premium Pay after Designated Hours[2]	
			Daily	Weekly
		$7.25		

The Delaware minimum wage is automatically replaced with the Federal minimum wage rate if it is higher than the State minimum.

District of Columbia Minimum Wage Rates

DISTRICT OF COLUMBIA	Future Effective Date	Basic Minimum Rate (per hour)	Premium Pay after Designated Hours[2]	
			Daily	Weekly
		$8.25		40

In the District of Columbia, the rate is automatically set at $1 above the Federal minimum wage rate if the District of Columbia rate is lower.

Florida Minimum Wage Rates

FLORIDA	Future Effective Date	Basic Minimum Rate (per hour)	Premium Pay after Designated Hours[2]	
			Daily	Weekly
		$7.25		N/A

Rate is increased annually based upon a cost of living formula.

Georgia Minimum Wage Rates

GEORGIA	Future Effective Date	Basic Minimum Rate (per hour)	Premium Pay after Designated Hours[2]	
			Daily	Weekly
(Applicable to employers of 6 or more employees)		$5.15		

The State law excludes from coverage any employment that is subject to the Federal Fair Labor Standards Act when the Federal rate is greater than the State rate.

Guam Minimum Wage Rates

GUAM	Future Effective Date	Basic Minimum Rate (per hour)	Premium Pay after Designated Hours[2]	
			Daily	Weekly
		$7.25		40

Hawaii Minimum Wage Rates

HAWAII	Future Effective Date	Basic Minimum Rate (per hour)	Premium Pay after Designated Hours[2]	
			Daily	Weekly
		$7.25		40

An employee earning a guaranteed monthly compensation of $2,000 or more is exempt from the State minimum wage and overtime law.

The State law excludes from coverage any employment that is subject to the Federal Fair Labor Standards Act unless the State wage rate is higher than the Federal.

Idaho Minimum Wage Rates

IDAHO	Future Effective Date	Basic Minimum Rate (per hour)	Premium Pay after Designated Hours[2]	
			Daily	Weekly
		$7.25		

Illinois Minimum Wage Rates

ILLINOIS	Future Effective Date	Basic Minimum Rate (per hour)	Premium Pay after Designated Hours[2]	
			Daily	Weekly
(Applicable to employers of 4 or more employees, excluding family members)		$8.25		40

Indiana Minimum Wage Rates

INDIANA	Future Effective Date	Basic Minimum Rate (per hour)	Premium Pay after Designated Hours[2]	
			Daily	Weekly
(Applicable to employers of 2 or more employees)		$7.25		40

Iowa Minimum Wage Rates

IOWA	Future Effective Date	Basic Minimum Rate (per hour)	Premium Pay after Designated Hours[2]	
			Daily	Weekly
		$7.25		

The Iowa minimum wage is automatically replaced with the Federal minimum wage rate if it is higher than the State minimum.

Kansas Minimum Wage Rates

KANSAS	Future Effective Date	Basic Minimum Rate (per hour)	Premium Pay after Designated Hours[2]	
			Daily	Weekly
				46
		$7.25		

The State law excludes from coverage any employment that is subject to the Federal Fair Labor Standards Act.

Kentucky Minimum Wage Rates

KENTUCKY	Future Effective Date	Basic Minimum Rate (per hour)	Premium Pay after Designated Hours[2]	
			Daily	Weekly
		$7.25		40
				7th day

The 7th day overtime law, which is separate from the minimum wage law differs in coverage from that in the minimum wage law and requires premium pay on the seventh day for those employees who work seven days in any one workweek.

The state adopts the Federal minimum wage rate by reference if the Federal rate is greater than the State rate.

Compensating time in lieu of overtime is allowed upon written request by an employee of any county, charter county, consolidated local government, or urban-county government, including an employee of a county-elected official.

Louisiana Minimum Wage Rates

LOUISIANA	Future Effective Date	Basic Minimum Rate (per hour)	Premium Pay after Designated Hours[2]	
			Daily	Weekly
There is no state minimum wage law		N/A		N/A

Maine Minimum Wage Rates

MAINE	Future Effective Date	Basic Minimum Rate (per hour)	Premium Pay after Designated Hours[2]	
			Daily	Weekly
		$7.50		40

The Maine minimum wage is automatically replaced with the Federal minimum wage rate if it is higher than the State minimum with the exception that any such increase is limited to no more than $1.00 per hour above the current legislated State rate.

Maryland Minimum Wage Rates

MARYLAND	Future Effective Date	Basic Minimum Rate (per hour)	Premium Pay after Designated Hours[2]	
			Daily	Weekly
		$7.25		40

The Maryland minimum wage is automatically replaced with the Federal minimum wage rate if it is higher than the State minimum wage rate.

Massachusetts Minimum Wage Rates

MASSACHUSETTS	Future Effective Date	Basic Minimum Rate (per hour)	Premium Pay after Designated Hours[2]	
			Daily	Weekly
		$8.00		40

The Massachusetts minimum wage rate automatically increases to 10 cents above the rate set in the Fair Labor Standards Act if the Federal minimum wage equals or becomes higher than the State minimum.

Michigan Minimum Wage Rates

MICHIGAN	Future Effective Date	Basic Minimum Rate (per hour)	Premium Pay after Designated Hours[2]	
			Daily	Weekly
(Applicable to employers of 2 or more employees)		$7.40		40

The State law excludes from coverage any employment that is subject to the Federal Fair Labor Standards Act unless the State wage rate is higher than the Federal.

Minnesota Minimum Wage Rates

MINNESOTA	Future Effective Date	Basic Minimum Rate (per hour)	Premium Pay after Designated Hours[2]	
			Daily	Weekly
Large employer (enterprise with annual receipts of $625,000 or more)		$6.15		48
Small employer (enterprise with annual receipts of less than $625,000)		$5.25		48

Mississippi Minimum Wage Rates

MISSISSIPPI	Future Effective Date	Basic Minimum Rate (per hour)	Premium Pay after Designated Hours[2]	
			Daily	Weekly
No state minimum wage law		N/A		N/A

Missouri Minimum Wage Rates

MISSOURI	Future Effective Date	Basic Minimum Rate (per hour)	Premium Pay after Designated Hours[2]	
			Daily	Weekly
		$7.25		40

In addition to the exemption for federally covered employment, the law exempts, among others, employees of a retail or service business with gross annual sales or business done of less than $500,000.

Premium pay required after 52 hours in seasonal amusement or recreation businesses.

Minimum wage is to be increased or decreased by a cost of living factor starting January 1, 2008 and every January 1 thereafter.

Montana Minimum Wage Rates

MONTANA	Future Effective Date	Basic Minimum Rate (per hour)	Premium Pay after Designated Hours[2]	
			Daily	Weekly
		$7.25		
State Law				
Except businesses with gross annual sales of $110,000 or less		$4.00		40

Minimum wage is subject to a cost of living adjustment done by September 30 of each year and effective on January 1 of the following year.

Nebraska Minimum Wage Rates

NEBRASKA	Future Effective Date	Basic Minimum Rate (per hour)	Premium Pay after Designated Hours[2]	
			Daily	Weekly
(Applicable to employers of 4 or more employees)		$7.25		

Nevada Minimum Wage Rates

NEVADA	Future Effective Date	Basic Minimum Rate (per hour)	Premium Pay after Designated Hours[2]	
			Daily	Weekly
		$8.25 (with no health ins. benefits provided by employer)	8	40
		$7.25 (with health ins. benefits provided by employer and received by employee)		

The premium overtime pay requirement on either a daily or weekly basis is not applicable to employees who are compensated at not less than one and one-half times the minimum rate or to employees of enterprises having a gross annual sales volume of less than $250,000.

The basic hourly rate is increased to $6.55 when the employer offers the employee a qualified health plan.

The minimum wage rate may be increased annually based upon changes in the cost of living index increase.

New Hampshire Minimum Wage Rates

NEW HAMPSHIRE	Future Effective Date	Basic Minimum Rate (per hour)	Premium Pay after Designated Hours[2]	
			Daily	Weekly
		$7.25		40

The New Hampshire minimum wage is automatically replaced with the Federal minimum wage rate if it is higher than the State minimum.

New Jersey Minimum Wage Rates

NEW JERSEY	Future Effective Date	Basic Minimum Rate (per hour)	Premium Pay after Designated Hours[2]	
			Daily	Weekly
		$7.25		40

New Mexico Minimum Wage Rates

NEW MEXICO	Future Effective Date	Basic Minimum Rate (per hour)	Premium Pay after Designated Hours[2]	
			Daily	Weekly
		$7.50		40

New York Minimum Wage Rates

NEW YORK	Future Effective Date	Basic Minimum Rate (per hour)	Premium Pay after Designated Hours[2]	
			Daily	Weekly
		$7.25		40

The New York minimum wage is automatically replaced with the Federal minimum wage rate if it is higher than the State minimum.

North Carolina Minimum Wage Rates

NORTH CAROLINA	Future Effective Date	Basic Minimum Rate (per hour)	Premium Pay after Designated Hours[2]	
			Daily	Weekly
		$7.25		40

Premium pay is required after 45 hours a week in seasonal amusements or recreational establishments.

North Dakota Minimum Wage Rates

NORTH DAKOTA	Future Effective Date	Basic Minimum Rate (per hour)	Premium Pay after Designated Hours[2]	
			Daily	Weekly
		$7.25		40

Ohio Minimum Wage Rates

OHIO	Future Effective Date	Basic Minimum Rate (per hour)	Premium Pay after Designated Hours[2]	
			Daily	Weekly
State Law		$7.30		40
		$7.25 (for those employers grossing $267,000 or less)		

Oklahoma Minimum Wage Rates

OKLAHOMA	Future Effective Date	Basic Minimum Rate (per hour)	Premium Pay after Designated Hours[2]	
			Daily	Weekly
Employers of 10 or more full-time employees at any one location and employers with annual gross sales over $100,000 irrespective of number of full time employees		$7.25		
All other employers		$2.00		

The Oklahoma state minimum wage law does not contain current dollar minimums. Instead the state adopts the Federal minimum wage rate by reference.

The State law excludes from coverage any employment that is subject to the Federal Fair Labor Standards Act.

Oregon Minimum Wage Rates

OREGON	Future Effective Date	Basic Minimum Rate (per hour)	Premium Pay after Designated Hours[2]	
			Daily	Weekly
		$8.40		40

Premium pay required after 10 hours a day in nonfarm canneries, driers, or packing plants and in mills, factories or manufacturing establishments (excluding sawmills, planning mills, shingle mills, and logging camps).

Beginning January 1, 2004, and annually thereafter, the rate will be adjusted for inflation by a calculation using the U.S. City Average Consumer Price Index for All Urban Consumers for All Items. The wage amount established will be rounded to the nearest five cents.

Pennsylvania Minimum Wage Rates

PENNSYLVANIA	Future Effective Date	Basic Minimum Rate (per hour)	Premium Pay after Designated Hours[2]	
			Daily	Weekly
		$7.25		40

Puerto Rico Minimum Wage Rates

PUERTO RICO	Future Effective Date	Basic Minimum Rate (per hour)	Premium Pay after Designated Hours[2]	
			Daily	Weekly
		$4.10	8	40 (double time)
			And on statutory rest day (double time)	

Employers covered by the Federal Fair Labor Standards Act (FLSA) are subject only to the Federal minimum wage and all applicable regulations. Employers not covered by the FLSA will be subject to a minimum wage that is at least 70 percent of the Federal minimum wage or the applicable mandatory decree rate, whichever is higher. The Secretary of Labor and Human Resources may authorize a rate based on a lower percentage for any employer who can show that implementation of the 70 percent rate would substantially curtail employment in that business.

Puerto Rico also has minimum wage rates that vary according to the industry. These rates range from a minimum of $4.25 to $7.25 per hour.

Rhode Island Minimum Wage Rates

RHODE ISLAND	Future Effective Date	Basic Minimum Rate (per hour)	Premium Pay after Designated Hours[2]	
			Daily	Weekly
		$7.40		40

Time and one-half premium pay for work on Sundays and holidays in retail and certain other businesses is required under two laws that are separate from the minimum wage law.

South Carolina Minimum Wage Rates

SOUTH CAROLINA	Future Effective Date	Basic Minimum Rate (per hour)	Premium Pay after Designated Hours[2]	
			Daily	Weekly
No state minimum wage law		N/A		N/A

South Dakota Minimum Wage Rates

SOUTH DAKOTA	Future Effective Date	Basic Minimum Rate (per hour)	Premium Pay after Designated Hours[2]	
			Daily	Weekly
		$7.25		

Tennessee Minimum Wage Rates

TENNESSEE	Future Effective Date	Basic Minimum Rate (per hour)	Premium Pay after Designated Hours[2]	
			Daily	Weekly
No state minimum wage law		N/A		N/A

The state does have a promised wage law whereby the employers are responsible for paying to the employees the wages promised by the employer.

Texas Minimum Wage Rates

TEXAS	Future Effective Date	Basic Minimum Rate (per hour)	Premium Pay after Designated Hours[2]	
			Daily	Weekly
		$7.25		

The State law excludes from coverage any employment that is subject to the Federal Fair Labor Standards Act.

The Texas State minimum wage law does not contain current dollar minimums. Instead the State adopts the Federal minimum wage rate by reference.

Utah Minimum Wage Rates

UTAH	Future Effective Date	Basic Minimum Rate (per hour)	Premium Pay after Designated Hours[2]	
			Daily	Weekly
		$7.25		

The Utah state minimum wage law does not contain current dollar minimums. Instead the state law authorizes the adoption of the Federal minimum wage rate via administrative action.

The State law excludes from coverage any employment that is subject to the Federal Fair Labor Standards Act.

Vermont Minimum Wage Rates

VERMONT	Future Effective Date	Basic Minimum Rate (per hour)	Premium Pay after Designated Hours[2]	
			Daily	Weekly
(Applicable to employers of two or more employees)		$8.06		40

The state overtime pay provision has very limited application because it exempts numerous types of establishments, such as retail and service; seasonal amusement/ recreation; hotels, motels, restaurants; and transportation employees to whom the Federal (FLSA) overtime provision does not apply.

The Vermont minimum wage is automatically replaced with the Federal minimum wage rate if it is higher than the State minimum.

Beginning January 1, 2007, and on each subsequent January 1, the minimum wage rate shall be increased by five percent or the percentage increase of the Consumer Price Index, or city average, not seasonally adjusted.

Virginia Minimum Wage Rates

VIRGINIA	Future Effective Date	Basic Minimum Rate (per hour)	Premium Pay after Designated Hours[2]	
			Daily	Weekly
(Applicable to employers of 4 or more employees)		$7.25		

The Virginia state minimum wage law does not contain current dollar minimums. Instead the state adopts the Federal minimum wage rate by reference.

The State law excludes from coverage any employment that is subject to the Federal Fair Labor Standards Act.

Virgin Islands Minimum Wage Rates

VIRGIN ISLANDS	Future Effective Date	Basic Minimum Rate (per hour)	Premium Pay after Designated Hours[2]	
			Daily	Weekly
State law		$7.25	8	40
				On 6th and 7th consecutive days.
Except businesses with gross annual receipts of less than $150,000		$4.30		

Washington Minimum Wage Rates

WASHINGTON	Future Effective Date	Basic Minimum Rate (per hour)	Premium Pay after Designated Hours[2]	
			Daily	Weekly
		$8.55		

Premium pay not applicable to employees who request compensating time off in lieu of premium pay.

Beginning January 1, 2001, and annually thereafter, the rate will be adjusted for inflation by a calculation using the consumer price index for urban wage earners and clerical workers for the prior year.

West Virginia Minimum Wage Rates

WEST VIRGINIA	Future Effective Date	Basic Minimum Rate (per hour)	Premium Pay after Designated Hours[2]	
			Daily	Weekly
(Applicable to employers of 6 or more employees at one location)		$7.25		40

Wisconsin Minimum Wage Rates

WISCONSIN	Future Effective Date	Basic Minimum Rate (per hour)	Premium Pay after Designated Hours[2]	
			Daily	Weekly
		$7.25		40

Wyoming Minimum Wage Rates

WYOMING	Future Effective Date	Basic Minimum Rate (per hour)	Premium Pay after Designated Hours[2]	
			Daily	Weekly
		$5.15		

[1] Like the Federal wage and hour law, State law often exempts particular occupations or industries from the minimum labor standard generally applied to covered employment. Particular exemptions are not identified in this table. Users are encouraged to consult the laws of particular States in determining whether the State's minimum wage applies to a particular employment. This information often may be found at the websites maintained by State labor departments. Links to these websites are available at www.dol.gov/whd/contacts/state_of.htm.

[2] The overtime premium rate is one and one-half times the employee's regular rate, unless otherwise specified.

This document was last revised in July 2010.

Consolidated State Minimum Wage Update Table (Effective Date: 07/01/2010)

> Federal MW	Equals Federal MW of $7.25	< Federal MW	No MW Required
AK - 7.75	AZ	AR - 6.25	AL
CA - 8.00	DE	CO - 7.24	LA
CT - 8.25	FL	GA - 5.15	MS
DC - 8.25	HI	MN - 6.15	SC
IL - 8.25	IA	WY - 5.15	TN
MA - 8.00	ID		
ME - 7.50	IN		
MI - 7.40	KS	5 States	5 States
NV - 8.25	KY		
NM - 7.50	MD		
OH - 7.30	MO		
OR - 8.40	MT		
RI - 7.40	NE		
VT - 8.06	NH		
WA - 8.55	NJ		
	NY		
	NC		
14 States + DC	ND		
	OK		
	PA		
	SD		
	TX		
	UT		
	VA		
	WV		
	WI		
	26 States		

Appendix 4: Spotlight on Statistics

July 2008

OLDER WORKERS

ARE THERE MORE OLDER PEOPLE IN THE WORKPLACE?

Between 1977 and 2007, employment of workers 65 and over increased 101 percent, compared to a much smaller increase of 59 percent for total employment (16 and over). The number of employed men 65 and over rose 75 percent, but employment of women 65 and older increased by nearly twice as much, climbing 147 percent. While the number of employed people age 75 and over is relatively small (0.8 percent of the employed in 2007), this group had the most dramatic gain, increasing 172 percent between 1977 and 2007 (Figure A4.1).

DOES THIS INCREASE JUST REFLECT THE AGING OF THE BABY-BOOM POPULATION?

No, because in 2007 the baby-boom generation — those individuals born between 1946 and 1964 — had not yet reached the age of 65.

Between 1977 and 2007, the age 65 and older civilian noninstitutional population — which excludes people in nursing homes — increased by about 60 percent, somewhat faster than the civilian noninstitutional population age 16 and over (46 percent). Yet employment of people 65 and over doubled while employment for everyone 16 and over increased by less than 60 percent. How can employment increase more than the population? A larger share of people 65 and older is staying in or returning to the labor force (which consists of those working or looking for work). The labor force participation rate for older workers has been rising since the late 1990s. This is especially notable because the 65-and-over labor force participation rate had been at historic lows during the 1980s and early 1990s (Figure A4.2).

ARE OLDER WORKERS CHOOSING PART-TIME OR FULL-TIME EMPLOYMENT?

Since the mid-1990s there has been a dramatic shift in the part-time versus full-time status of the older workforce. The ratio of part-time to full-time employment among older workers was relatively steady from 1977 through 1990. Between 1990 and 1995, part-time work among older workers began trending upward with a corresponding decline in full-time employment. But after 1995, that trend began a marked reversal with full-time employment rising sharply. Between 1995 and 2007,

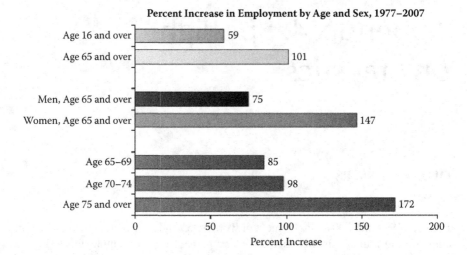

FIGURE A4.1   (www.bls.gov/spotlight/2008/Older_workers/data.htm#chart_01)

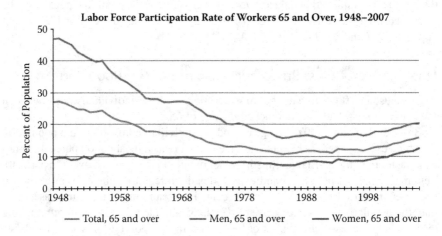

FIGURE A4.2   (www.bls.gov/spotlight/2008/Older_workers/data.htm#chart_02)

the number of older workers on full-time work schedules nearly doubled while the number working part-time rose just 19 percent. As a result, full-timers now account for a majority among older workers: 56 percent in 2007, up from 44 percent in 1995 (Figure A4.3).

WHAT PORTION OF EMPLOYED OLDER WOMEN ARE MARRIED?

In 1977, about one-third of employed women 65 and older were married, but by 2007, married women accounted for nearly one-half of these workers. Women workers who were widowed, divorced or separated represented 56 percent of employed women 65 and older in 1977; by 2007 their share had fallen to 48 percent. During

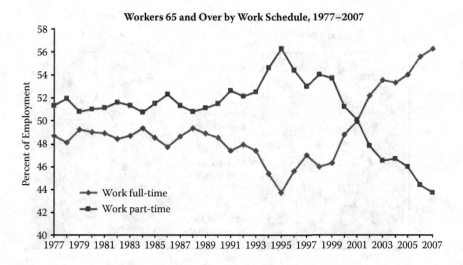

FIGURE A4.3   (www.bls.gov/spotlight/2008/Older_workers/data.htm#chart_03)

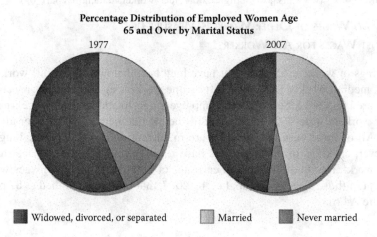

FIGURE A4.4   (www.bls.gov/spotlight/2008/Older_workers/data.htm#chart_04)

the same time period, the fraction of older women workers who were never married shrank from about 11 percent to about 6 percent (Figure A4.4).

How do Older Workers Stack Up against Younger Workers in Terms of Education?

It wasn't that long ago that older and younger workers had very different educational backgrounds. In 1997, 21 percent of employed older workers had less than a high school education compared to only 10 percent of those ages 25-64. By 2007, older workers with less than a high school education accounted for just 13 percent of that group's employment, compared with 9 percent for younger workers (Figure A4.5).

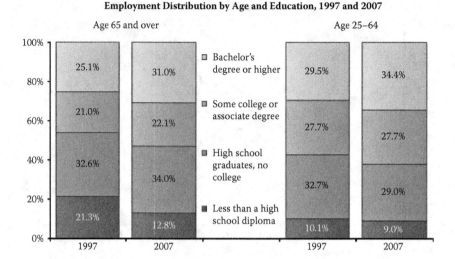

Employment Distribution by Age and Education, 1997 and 2007

FIGURE A4.5   (www.bls.gov/spotlight/2008/Older_workers/data.htm#chart_05)

How do Wages of Older Workers Measure up against Wages for All Workers?

Earnings of workers 65 and older have long been below those of all workers. In 1979, median weekly earnings for full-time workers age 65 and older were $198 compared to $240 for all full-time employees age 16 and up. In 2007, earnings of older workers were $605 per week, still below the median of $695 for all workers. (All of these earnings amounts are in current dollars.) Over the long term, however, earnings of older workers have risen at a slightly faster pace than the total workforce. In 1979, median earnings of older full-time employees were 83 percent of those ages 16 and up; but, by 2007, that ratio had climbed to 87 percent (Figure A4.6).

How Does Inflation Affect Older Workers?

A number of years ago, the Bureau of Labor Statistics created an experimental consumer price index (CPI) for Americans 62 years of age and older. In this index, items purchased more frequently by the older population, such as medical care, have a higher weight than in the official CPI (which covers a much broader share of the population); items purchased less frequently, such as clothing, have a lower weight. Data from the experimental series show that the annual inflation rate for seniors has been equal to or greater than the inflation rate for all urban consumers in every year since that series began except for 1983 and 2007. However, the yearly differences have been fairly small; over the past 25 years the index for older Americans has risen an average of 3.3 percent each year, as compared to 3.1 percent for the official CPI (Figure A4.7).

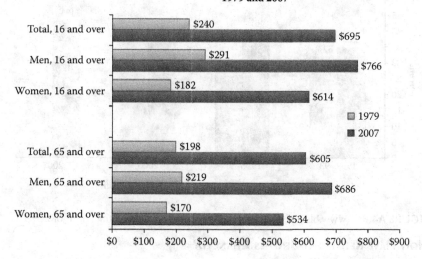

FIGURE A4.6   (www.bls.gov/spotlight/2008/0lder_workers/data.htm#chart_06)

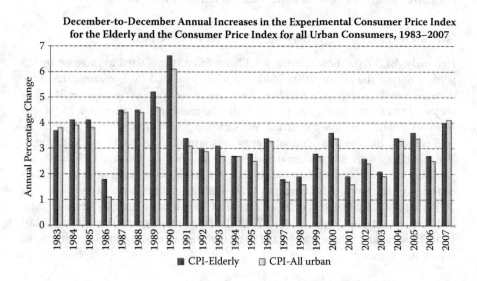

FIGURE A4.7   (www.bls.gov/spotlight/2008/0lder_workers/data.htm#chart_07)

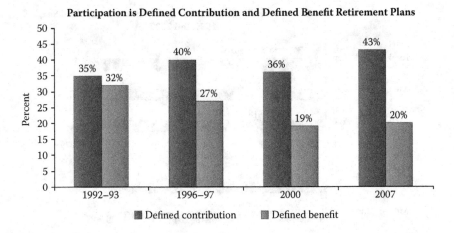

FIGURE A4.8   (www.bls.gov/spotlight/2008/Older_workers/data.htm#chart_08)

HOW HAVE RETIREMENT BENEFITS CHANGED?

Among all workers, participation in defined benefit plans has fallen while participation in defined contribution plans has risen. In defined benefit plans, companies
promise to pay workers a specified amount in retirement benefits. In defined contribution plans, companies promise to contribute a specified amount, but make no
assurance as to the final payout. Among all workers, there has been a decrease in
the percentage covered by defined benefit ("payout") plans and an increase in the
percentage covered by defined contribution ("pay in") plans. For more and more
workers, this means that risk — in terms of steady retirement income — has been
transferred from the employer to the eventual retiree (Figure A4.8).

IS THIS GRAYING OF THE WORKFORCE EXPECTED TO CONTINUE?

Definitely. BLS data show that the total labor force is projected to increase by 8.5
percent during the period 2006-2016, but when analyzed by age categories, very
different trends emerge. The number of workers in the youngest group, age 16-24, is
projected to decline during the period while the number of workers age 25-54 will
rise only slightly. In sharp contrast, workers age 55-64 are expected to climb by
36.5 percent. But the most dramatic growth is projected for the two oldest groups.
The number of workers between the ages of 65 and 74 and those aged 75 and up are
predicted to soar by more than 80 percent. By 2016, workers age 65 and over are
expected to account for 6.1 percent of the total labor force, up sharply from their
2006 share of 3.6 percent. (For more data see Civilian labor force by sex, age, race,
and Hispanic origin (Figure A4.9).) With the baby-boom generation about to start
joining the ranks of those age 65 and over, the graying of the American workforce
is only just beginning.

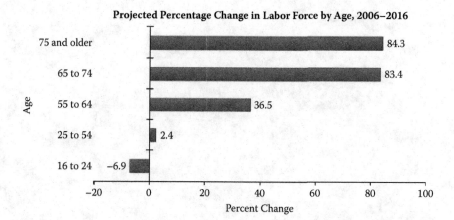

FIGURE A4.9   (www.bls.gov/spotlight/2008/Older_workers/data.htm#chart_09)

Appendix 5: OSHA Instruction

Directive Number: CPL 02-00-149	Effective Date: June 18, 2010
Subject: Severe Violator Enforcement Program (SVEP)	

ABSTRACT

Purpose: This Instruction establishes enforcement policies and procedures for OSHA's Severe Violator Enforcement Program (SVEP), which concentrates resources on inspecting employers who have demonstrated indifference to their OSH Act obligations by willful, repeated, or failure-to-abate violations. This Instruction replaces OSHA's Enhanced Enforcement Program (EEP).

Scope: OSHA-wide.

References: OSHA Instruction CPL 02-00-069, Special Emphasis: Trenching and Excavation, September 19, 1985; OSHA Instruction CPL 02-00-080, Handling of Cases To Be Proposed for Violation-By-Violation Penalties, October 21, 1990; OSHA Instruction CPL 02-00-136, National Emphasis Program (NEP) on Shipbreaking, March 16, 2005; OSHA Instruction CPL 02-00-148, Field Operations Manual (FOM), November 9, 2009; OSHA Instruction CPL 03-00-003, National Emphasis Program on Amputations, October 27, 2006; OSHA Instruction CPL 03-00-010, Petroleum Refinery Process Safety Management National Emphasis Program, August 18, 2009; OSHA Instruction CPL 03-00-007, National Emphasis Program— Crystalline Silica, January 24, 2008; OSHA Instruction CPL 03-00-008—Combustible Dust National Emphasis Program (Reissued), March 11, 2008; and OSHA Notice 09-06 (CPL 02) PSM Covered Chemical Facilities National Emphasis Program, July 27, 2009.

Cancellations: OSHA Instruction CPL 02-00-145, Enhanced Enforcement Program (EEP), January 1, 2008.

State Impact: Notice of Intent, Adoption and Submission of a Plan Change Required. See section VI.

Action Offices: National, Regional, Area Offices and State Plan States.

Originating Office: Directorate of Enforcement Programs.

Contact: Directorate of Enforcement Programs
Office of General Industry Enforcement
200 Constitution Avenue, NW, N-3119
Washington, DC 20210

By and Under the Authority of
David Michaels, PhD, MPH
Assistant

EXECUTIVE SUMMARY

This Instruction establishes enforcement policies and procedures for OSHA's Severe Violator Enforcement Program (SVEP), which concentrates resources on inspecting employers who have demonstrated indifference to their OSH Act obligations by committing willful, repeated, or failure-to-abate violations. Enforcement actions for severe violator cases include mandatory follow-up inspections, increased company/corporate awareness of OSHA enforcement, corporate-wide agreements, where appropriate, enhanced settlement provisions, and federal court enforcement under Section 11(b) of the OSH Act. In addition, this Instruction provides for nationwide referral procedures, which includes OSHA's State Plan States. This Instruction replaces OSHA's Enhanced Enforcement Program (EEP).

SIGNIFICANT CHANGES FROM THE ENHANCED ENFORCEMENT PROGRAM (EEP)

- High-Emphasis Hazards are targeted, which include fall hazards and specific hazards identified from selected National Emphasis Programs.
- The Assistant Secretary has determined that Nationwide Inspections of Related Workplaces/Worksites are critical inspections for the purpose of 29 CFR §1908.7(b)(2)(iv).
- Creates a nationwide referral procedure for Regions and State Plan States.

TABLE OF CONTENTS

  I. Purpose.

This Instruction establishes enforcement policies and procedures for OSHA's Severe Violator Enforcement Program (SVEP), which concentrates resources on inspecting employers who have demonstrated indifference to their OSH Act obligations by committing willful, repeated, or failure-to-abate violations. This Instruction replaces OSHA's Enhanced Enforcement Program (EEP).

  II. Scope.

This Instruction applies OSHA-wide.

III. References.
   A. OSHA Instruction CPL 02-00-069—CPL 2.69—Special Emphasis: Trenching and Excavation, September 19, 1985.
   B. OSHA Instruction CPL 02-00-136, National Emphasis Program (NEP) on Shipbreaking, March 16, 2005.
   C. OSHA Instruction CPL 02-00-148—Field Operations Manual (FOM), November 9, 2009.
   D. OSHA Instruction CPL 03-00-003—National Emphasis Program on Amputations, October 27, 2006.
   E. OSHA Instruction CPL 03-00-010—Petroleum Refinery Process Safety Management National Emphasis Program, August 18, 2009.
   F. OSHA Instruction CPL 03-00-007—National Emphasis Program— Crystalline Silica, January 24, 2008.
   G. OSHA Instruction CPL 03-00-008—Combustible Dust National Emphasis Program (Reissued), March 11, 2008.
   H. OSHA Instruction CPL 03-00-009—OSHA Instruction; National Emphasis Program—Lead, August 14, 2008.
   I. OSHA Instruction 09-06 (CPL 02)—PSM Covered Chemical Facilities National Emphasis Program, July 27, 2009.
   J. OSHA Instruction CSP 01-00-002, State Plan Policies and Procedures Manual, March 21, 2001.
   K. OSHA Instruction CPL 02-00-080 Handling of Cases To Be Proposed for Violation-By-Violation Penalties, October 21, 1990.
IV. Cancellations.
   OSHA Instruction CPL 02-00-145, Enhanced Enforcement Program (EEP), January 1, 2008.
V. Action Information.
   A. Responsible Office.
      Directorate of Enforcement Programs (DEP).
   B. Action Offices.
      National Office, Regional Offices, Area Offices, and State Plan States.
   C. Information Offices.
      OSHA Training Institute, Consultation Project Managers, VPP Managers and Coordinators, OSHA Strategic Partnership Coordinators, Compliance Assistance Coordinators, and Compliance Assistance Specialists.
VI. State Impact.
   Notice of Intent, Adoption and Submission of a Plan Change Supplement Required. This Instruction is a Federal program change which cancels OSHA's Enhanced Enforcement Program and establishes the Severe Violator Enforcement Program (SVEP). The purpose of the SVEP is to focus increased enforcement attention on significant hazards and violations by concentrating on employers who have demonstrated indifference to their occupational safety and health obligations through willful, repeated, or failure-to-abate violations in (1) a fatality or catastrophe situation; (2) in industry operations or processes that expose employees to the most severe occupational hazards and those identified as "High-Emphasis Hazards," as

defined in Section XII. of this Instruction (p. 5); (3) exposing employees to hazards related to the potential release of a highly hazardous chemical; or (4) all egregious enforcement actions. Because the significant nature of this program requires nationwide applicability, States are required to either adopt this program or establish their own equivalent program which must include enforcement procedures for identifying and taking action with regard to these recalcitrant and indifferent employers and for making referrals to, and responding to referrals from, Federal OSHA. As set out in paragraph XV.B.3. (p. 12), Federal OSHA will accept referrals from State plans and conduct appropriate inspections.

Each State must provide notice of its intent to adopt either policies or procedures identical to those set out in this instruction, or alternative policies and procedures that are at least as effective. State policies and procedures must be adopted within six months of issuance of this instruction. If the State adopts policies and procedures that differ from the Federal program, the State must submit its different policies and procedures as a plan change supplement with identification of the differences, within 60 days of adoption, and either post its different policies on its State plan website and provide the link to OSHA or provide OSHA with information on how the public may obtain a copy. If the State adopts identical policies and procedures, it must provide the date of adoption to OSHA and information on any State-specific procedures. OSHA will provide summary information on the State responses to this instruction on its website.

VII. Background.

The SVEP is intended to focus enforcement efforts on significant hazards and violations by concentrating inspection resources on employers who have demonstrated recalcitrance or indifference to their OSH Act obligations by committing willful, repeated, or failure-to-abate violations in one or more of the following circumstances: (1) a fatality or catastrophe situation; (2) in industry operations or processes that expose employees to the most severe occupational hazards and those identified as "High-Emphasis Hazards," as defined in Section XII of this Instruction (p.5); (3) exposing employees to hazards related to the potential release of a highly hazardous chemical; or (4) all egregious enforcement actions.

Cases meeting the severe violator enforcement criteria are those in which the employer is found to be recalcitrant or indifferent to its obligations under the OSH Act, thereby endangering employees. The SVEP procedures in Section XV. (p. 10) are intended to increase attention on the correction of hazards found in these workplaces and, where appropriate, in other worksites of the same employer where similar hazards and deficiencies may be present. This program applies to all employers regardless of size.

VIII. Transition between the EEP and the SVEP.

On the effective date of this Instruction, both the original EEP (referred to as EEP or EEP1) and the revised EEP (referred to as EEP2) will terminate (including any follow-up inspections that have not yet been performed). However, Area Directors have the discretion to conduct any follow-up inspections related to the EEP in accordance with the policies and procedures in the FOM (CPL 02-00-148, Chapter 2, *Program Planning*).

IX. Significant Changes.

A. High-Emphasis Hazards are targeted and include fall hazards and hazards identified from the following National Emphasis Programs (NEPs): amputations, combustible dust, crystalline silica, excavation/trenching, lead, and shipbreaking. See section XII. (p. 5).

B. The Assistant Secretary by this Instruction determines that the specific inspections under the Nationwide Inspections of Related Workplaces/Worksites process are critical inspections for the purpose of 29 CFR §1908.7(b)(2)(iv) and delegates to the Director of the Directorate of Enforcement Programs the authority to determine site selections of those related workplaces/worksites. See paragraph XV.B.7.b. (p. 16).

C. A nationwide referral procedure is being initiated in which OSHA may inspect related worksites/workplaces of a SVEP employer. See section XV.B. (p. 11).

X. Handling Severe Violator Enforcement Cases.

A. Compliance Officers (CSHOs) must become familiar with Appendix B to effectively evaluate employers during all inspections likely to result in a severe violator enforcement case.

B. The Area Director will identify severe violator enforcement cases no later than at the time the citations are issued, in accordance with criteria set forth in this Instruction.

C. Federal Agency cases that meet the SVEP case criteria will also be classified as severe violator enforcement cases, and where the term "employer-wide" or "company-wide" is used, it will apply agency-wide or department-wide, as appropriate. Appropriate SVEP actions for such cases will be determined by the Area Director in consultation with the Regional Administrator.

D. When a case meets the severe violator enforcement case criteria, the Area Director will notify the Regional Administrator, who in turn will notify the Directorate of Enforcement Programs (DEP).

E. Regional Administrator notification to DEP must be by e-mail using the SVEP-group e-mail address on OSHA's Global Address list: "OSHA-SVEP." The notification must be at least monthly (by the 20th of the month) and include the information requested in Appendix A. Regions must use the Excel spreadsheet format that will be sent to the SVEP Regional Coordinators shortly after this Instruction becomes effective.

XI. Criteria for a Severe Violator Enforcement Case.

Any inspection that meets one or more of criteria A. through D., at the time that the citations are issued, will be considered a severe violator enforcement case.

Willful and **repeated** citations and **failure-to-abate** notices must be based on serious violations, except for recordkeeping, which must be egregious (e.g., per-instance citations). See *FOM*, CPL 02-00-148, Chapter 6, paragraphs V.A.1. and VI.A.1.

A. Fatality/Catastrophe Criterion.

A fatality/catastrophe inspection in which OSHA finds **one** or more **willful** or **repeated** violations or **failure-to-abate** notices based on a serious violation related to a death of an employee or three or more hospitalizations.

NOTE: The violations under this criterion **do not have to be High-Emphasis Hazards** as defined in Section XII. (p. 5).

B. Non-Fatality/Catastrophe Criterion Related to High-Emphasis Hazards.

An inspection in which OSHA finds **two** or more **willful** or **repeated** violations or **failure-to-abate** notices (or any combination of these violations/notices), based on **high gravity serious** violations related to a **High-Emphasis Hazard** as defined in Section XII. (p. 5).

C. Non-Fatality/Catastrophe Criterion for Hazards Due to the Potential Release of a Highly Hazardous Chemical (Process Safety Management).

An inspection in which OSHA finds **three** or more **willful** or **repeated** violations or **failure-to-abate** notices (or any combination of these violations/notices), based on **high gravity serious** violations related to hazards due to the potential release of a highly hazardous chemical, as defined in the PSM standard.

D. Egregious Criterion.

All **egregious** (e.g., per-instance citations) enforcement actions will be considered SVEP cases.

XII. Definition of High-Emphasis Hazards.

High-Emphasis Hazards as used in this Instruction means only **high gravity serious** violations of the following specific standards covered under falls or the National Emphasis Programs (NEPs) listed in paragraphs F. through K. below, **regardless of the type of inspection** being conducted (e.g., complaint, SST, Local Emphasis Programs, National Emphasis Programs). Low and moderate gravity violations **will not** be considered for a SVEP case. See Chapter 4, section II, *Serious Violations*, and Chapter 6, section III.A *Gravity of Violation* of the FOM (CPL 02-00-148) for determining what constitutes a **high gravity serious** violation.

Example 1: A CSHO conducts an **SST inspection** and cites the employer for one high gravity willful violation of 29 CFR §1910.23 and one low gravity willful violation of 29 CFR §1910.28. The inspection has not met the Non-Fatality/Catastrophe Criterion Related to High-Emphasis Hazards and is not subject to the SVEP.

Example 2: A CSHO conducts a **Local Emphasis Program** inspection for Residential Construction. While on-site, the CSHO observes employees working in an unsupported trench and cites the employer for two high gravity willful violations of 29 CFR §1926.651. The inspection has met the Non-Fatality/Catastrophe Criterion Related to High-Emphasis Hazards and the case is subject to the SVEP.

Example 3: A CSHO conducts a **National Emphasis Program** inspection for Shipbreaking. While on-site, the CSHO observes a piece of scrap metal from the dismantled vessel being lifted with a crane over top of an employee with a kinked wire rope. The employer is cited for one high gravity repeat violation of 29 CFR §1915.112 and one high gravity willful violation of 29 CFR §1915.116. The inspection has met the Non-Fatality/Catastrophe Criterion Related to High-Emphasis Hazards and the case is subject to the SVEP.

A. Fall hazards covered under the following general industry standards:
  1. 29 CFR §1910.23—Guarding floor and wall openings and holes [Walking-Working Surfaces]
  2. 29 CFR §1910.28—Safety requirements for scaffolding [Walking-Working Surfaces]
  3. 29 CFR §1910.29—Manually propelled mobile ladder stands and scaffolds (towers) [Walking-Working Surfaces]
  4. 29 CFR §1910.66—Powered platforms for building maintenance [Powered Platforms, Manlifts, and Vehicle-Mounted Work Platforms]
  5. 29 CFR §1910.67—Vehicle-mounted elevating and rotating work platforms [Powered Platforms, Manlifts, and Vehicle-Mounted Work Platforms]
  6. 29 CFR §1910.68—Manlifts [Powered Platforms, Manlifts, and Vehicle-Mounted Work Platforms]

B. Fall hazards covered under the following construction industry standards:
  1. 29 CFR §1926.451—General requirements [Scaffolds]
  2. 29 CFR §1926.452—Additional requirements applicable to specific types of scaffolds
  3. 29 CFR §1926.453—Aerial lifts [Scaffolds]
  4. 29 CFR §1926.501—Duty to have fall protection
  5. 29 CFR §1926.502—Fall protection systems criteria and practices
  6. 29 CFR §1926.760—Fall protection [Steel Erection]
  7. 29 CFR §1926.1052—Stairways [Ladders]

C. Fall hazards covered under the following shipyard standards:
  1. 29 CFR §1915.71—Scaffolds or staging [Scaffolds, ladders and Other Working Surfaces]
  2. 29 CFR §1915.73—Guarding of deck openings and edges [Scaffolds, ladders and Other Working Surfaces]
  3. 29 CFR §1915.74—Access to vessels [Scaffolds, ladders and Other Working Surfaces]
  4. 29 CFR §1915.75—Access to and guarding of dry docks and marine railways [Scaffolds, ladders and Other Working Surfaces]
  5. 29 CFR §1915.159—Personal fall arrest systems (PFAS) [Personal Protective Equipment (PPE)]

D. Fall hazards covered under the following marine terminal standards:
  1. 29 CFR §1917.45—Cranes and derricks [Cargo Handling Gear and Equipment]
  2. 29 CFR §1917.49—Spouts, chutes, hoppers, bins, and associated equipment [Cargo Handling Gear and Equipment]
  3. 29 CFR §1917.112—Guarding of edges [Terminal Facilities]

E. Fall hazards covered under the following longshoring standards:

1. 29 CFR §1918.22—Gangways [Gangways and Other Means of Access]
2. 29 CFR §1918.85—Containerized cargo operations [Handling Cargo]

F. **Amputation hazards** specified below that are covered under the National Emphasis Program on Amputations. (See CPL03-00-003):
   1. 29 CFR §1910.147—The control of hazardous energy (lockout/tagout)
   2. 29 CFR §1910.212—General requirements for all machines
   3. 29 CFR §1910.213—Woodworking machinery requirements
   4. 29 CFR §1910.217—Mechanical power presses
   5. 29 CFR §1910.219—Mechanical power-transmission apparatus

G. **Combustible dust hazards** specified below that are covered by the Combustible Dust National Emphasis Program (Reissued), including the General Duty Clause (Sec. 5(a)(1) of the OSH Act). (See CPL 03-00-008):
   1. 29 CFR §1910.22—General requirements [Walking-Working Surfaces]
   2. 29 CFR §1910.307—Hazardous (classified) locations [Electrical]
   3. Sec. 5(a)(1) of the OSH Act
   Any General Duty Clause violation concerning hazards related to dust collectors inside buildings, deflagration isolation systems, and ductwork issues.

H. **Crystalline silica hazards** specified below that are covered by the National Emphasis Program—Crystalline Silica (See CPL 03-00-007):
   1. Overexposure.
      a. 29 CFR Part §1910.1000 and 29 CFR Part §1915.1000—Air Contaminants
      b. 29 CFR §1926.55—Gases, vapors, fumes, dusts, and mists
   2. Failure to Implement Engineering Controls.
      a. 29 CFR §1910.1000(e)—Air Contaminants
      b. 29 CFR §1926.55(b)—Gases, vapors, fumes, dusts, and mists
   3. When Overexposure Occurs.
      29 CFR §1910.134; 29 CFR §1926.103; and 29 CFR §1915.154—Respiratory protection
      NOTE: The Silica NEP requires a mandatory follow-up inspection when overexposures to crystalline silica are found. If a follow-up inspection finds the same violations as previously cited, the follow-up inspection will most likely qualify as a SVEP case. See paragraph XV.A.4. (p. 11).

I. **Lead hazards** specified below that are covered by the National Emphasis Program—Lead (only violations based on sampling). See CPL 03-00-009.
   1. 29 CFR §1910.1025— Lead
   2. 29 CFR §1926.62—Lead
   3. 29 CFR §1915.1025—Lead and 29 CFR §1915 Subpart D Welding, Cutting, and Heating

J. **Excavation/trenching hazards** specified below that are covered by
the Special Emphasis Program—Trenching and Excavation (See CPL
02-00-069)
1. 29 CFR §1926.651—Specific excavation requirements
2. 29 CFR §1926.652—equirements for protective systems [Excavations]
K. **Shipbreaking hazards** specified below that are covered by the National
Emphasis Program—Shipbreaking. See CPL 02-00-136.
1. 29 CFR §1915.12—Precautions and the order of testing before entering confined and enclosed spaces and other dangerous atmospheres
[Confined and Enclosed Spaces and Other Dangerous Atmospheres
in Shipyard Employment]
2. 29 CFR §1915.112—Ropes, chains, and slings [Gear and Equipment
for Rigging and Materials Handling]
3. 29 CFR §1915.116—Use of Gear [Gear and Equipment for Rigging
and Materials Handling]
4. 29 CFR §1915.159—Personal fall arrest systems (PFAS) [Personal
Protective Equipment (PPE)]
5. 29 CFR §1915.503—Precautions for hot work [Fire Protection in
Shipyard Employment]

XIII. Hazards Due to the Potential Release of a Highly Hazardous Chemical
(Process Safety Management).
Petroleum refinery hazards are those hazards that are covered by the
Petroleum Refinery Process Safety Management National Emphasis
Program (See CPL 03-00-004) and hazards due to the potential release
of a highly hazardous chemical as covered by the PSM Covered Chemical
Facilities National Emphasis Program See Instruction 09-06 (CPL 02):
29 CFR §1910.119—Process safety management of highly hazardous
chemicals
XIV. Enforcement Considerations—Two or More Inspections of the Same
Employer.
For classification under the SVEP, each individual inspection must be eval-
uated separately to determine if it meets one of the criteria in XI.A., B., C.,
or D. (p. 4). If any of the inspections meet one of the severe violator crite-
ria, it will be considered a SVEP case and coded according to paragraph
XVIII.A. (p. 20).
XV. Procedures of the Severe Violator Enforcement Program (SVEP).
When the Area Director determines that a case meets one of the SVEP
criteria, it will be treated in accordance with paragraphs XV.A. through
E below. Only those SVEP actions that are appropriate for the particular
employer should be taken.
A. Enhanced Follow-up Inspections.
1. General.
For any SVEP inspection issued on or after the effective date of this
Instruction, a follow-up inspection must be conducted after the cita-
tions become final orders even if abatement verification of the cited

violations has been received. The purpose of the follow-up inspection is to assess **not only** whether the cited violation(s) were abated, **but also** whether the employer is committing similar violations.

2. Compelling Reason Not to Conduct.

    If there is a compelling reason not to conduct a follow-up inspection, the reason must be documented in the file. **The Region shall also report these cases monthly to the Director of Enforcement Programs**, along with the reason a follow-up was not initiated.

    If a follow-up cannot be initiated, the follow-up column of the SVEP Log must be completed by giving the reason. Examples of compelling reasons not to conduct a follow-up inspection may include: (1) worksite/workplace closed, (2) employer is out of business, (3) operation cited has been discontinued at the worksite/workplace, or (4) case no longer meets any of the SVEP criteria because citation has been withdrawn/vacated.

    NOTE: A Corrected During Inspection (CDI) situation does not take the place of a needed follow-up inspection.

    If the Area Director learns that a cited operation has been moved from the cited location to a different location, the new location shall be inspected. If the new location is outside the area office jurisdiction, a referral shall be made.

3. Construction Worksites.

    When the Area Office has reason to believe that a construction worksite is no longer active (or is nearing completion), thus making a follow-up inspection of the same worksite impossible or impractical, the provisions in paragraph XV.B.5. (p. 15) shall apply. When a construction follow-up is attempted but the employer is no longer at the site, it will not be added to the SVEP Log.

4. Silica Overexposure Follow-ups.

    The Silica NEP (CPL 03-00-007) in paragraph XI.E.1. requires a mandatory follow-up inspection when citations are issued for overexposure to crystalline silica to determine whether the employer is eliminating silica exposures or reducing exposures below the PEL. If a follow-up inspection finds the same or similar violations as previously cited, the follow-up inspection will most likely qualify as a severe violator enforcement case under the criteria in section XI. (p. 4).

B. Nationwide Inspections of Related Workplaces/Worksites.

1. General.

    OSHA has found that employer indifference to compliance responsibilities under the Act may be indicative of broader patterns of non-compliance at related employer worksites. When there are reasonable grounds to believe that compliance problems identified in the initial inspection may be indicative of a broader pattern of non-compliance, OSHA will inspect related worksites of the same employer. Appendix B of this Instruction provides guidance in evaluating whether compliance problems found during the initial

SVEP inspection are localized or likely to exist at related facilities. This information should be gathered, to the extent possible, during the initial SVEP inspection. Such information may also be sought by letter, telephone, or if necessary, by subpoena.

The Regional Administrator shall be responsible for assuring that relevant information is gathered and for determining whether the information provides reasonable grounds to believe that a broader pattern of non-compliance may exist. The Regional Administrator shall consult with the Regional Solicitor as appropriate. When sufficient evidence is found, all related establishments of the employer that are in the same 3-digit NAICS code (or 2-digit SIC code) as the initial SVEP case will be identified; establishments will be selected for inspection in accordance with subsection 4 below. Establishments that are not in the same 3-digit NAICS code (or 2-digit SIC code) also may be inspected if there are reasonable grounds to believe hazards and violations may be present at the related sites.

The Directorate of Enforcement programs will serve as the National Office point of contact for all SVEP nationwide referrals. Any questions should be addressed to the Director or Deputy Director in DEP. All Regional Administrators will name a SVEP Coordinator.

2. Office of Statistical Analysis (OSA).

At the request of the Director of the Directorate of Enforcement Programs, the Regional Administrator, or the Regional Coordinator, OSA will assist in identifying similar and other related worksites nationwide (including in State Plan States) of the same employer.

Establishments are related when there is common ownership. Related establishments include establishments of corporations that are in the same corporate family, including subsidiary, affiliate, or parent corporations with substantial common ownership.

Similar related establishments are related establishments that are in the same 3-digit NAICS code (or 2-digit SIC code).

3. State Plan State Referrals.

OSHA will accept referrals, which include all relevant facts, from State Plan States regarding any inspections conducted pursuant to the State's SVEP. State Plan referrals to Federal OSHA are to be sent to the Regional Administrator, who will forward any referrals not in its Region to the appropriate OSHA Regional Administrator.

4. General Industry Workplaces.

a. Employer Has Three (3) or Fewer Similar Related Workplaces. When the Regional Administrator determines that additional workplaces within that region should be inspected, and the employer has three or fewer similar related workplaces, **all such workplaces will be inspected** to determine whether those sites have hazardous conditions or violations similar to those in the severe violator enforcement case. The Regional Administrator shall have overall responsibility for coordinating the inspections and planning investigative strategy. The Regional Administrator shall consult with the Regional Solicitor as appropriate.

When any of the three or fewer workplaces that the Regional Administrator believes should be inspected are in **one or more of the Region's State Plan States**, the information will be forwarded to the State Plan Designee for inspection. A copy of the referral will also be sent to the Director of DEP.

When any of the three or fewer workplaces that the Regional Administrator believes should be inspected are in **two or more Regions or a State Plan State in another Region**, the information will be forwarded to the appropriate Regional Administrator for inspection. A copy of the referral will also be sent to the Director of DEP.

b.   Employer Has Four (4) or More Similar Related Workplaces. When the Regional Administrator determines that additional workplaces should be inspected, and the employer has **four or more similar related establishments within the Region or in other Regions**, or the number of workplaces/worksites cannot be determined, the Regional Administrator will send the recommendation for inspection, including all relevant facts, to the Director of DEP. The Director shall consult with the Associate Solicitor as appropriate.

(1)   When the Director of DEP determines that there are reasonable grounds for inspecting similar related establishments, he/she will issue a SVEP Nationwide inspection list. Normally, when the number of similar related establishments nationwide is 10 or less, all will be selected for inspection. When there are more than 10, the Office of Statistical Analysis will assign random numbers to the complete list of similar related establishments, sort those establishments in random number order, and select the first 10 for inspection.

All establishments on the inspection list will be inspected to determine whether hazardous conditions or violations similar to those found in the initial SVEP inspection are present. Based on the results of these inspections, the Director may determine whether inspections of additional establishments should be conducted. Any inspection conducted under a SVEP Nationwide inspection list is to be coded as an unprogrammed-referral, and is to be considered a referral from the National Office. An OSHA-90 is to be generated when a site is discovered where a SVEP Nationwide Referral employer is working.

(2)   In addition to or in lieu of (1) above, when the Director has reasonable grounds to believe that hazards may exist at par-

ticular other related establishments, he/she may select those establishments for inspection.

(3)   The Director shall be responsible for coordinating nationwide inspections of related establishments under this paragraph. Where complex or systemic issues are present, the Director shall convene a team to advise on investigative strategies, such as the use of administrative depositions or experts, and share information among offices participating in the inspections. The team shall include representatives from the OSHA and SOL national offices and regional offices where inspections will be conducted. In the event the inspections result in multiple contested citations, the team will advise SOL on litigation strategies that take account of such matters as distribution of work among affected offices and budget.

c.   SVEP Nationwide Related Inspections that involve process safety management (PSM) hazards.

A SVEP Nationwide inspection will be limited to investigations of the PSM standard for which the willful or repeated citations or failure-to-abate notices were issued, and will not include units that were inspected in the previous two years.

5. Construction Worksites.

a.   Regional Office.

Whenever an employer in the construction industry has a SVEP case, the Regional Administrator must further investigate the employer's OSH Act compliance. If the initially inspected worksite is closed before a follow-up inspection can be conducted, at least one other worksite of the cited employer must be inspected to determine whether the employer is committing violations similar to those found in the initial severe violator enforcement inspection. Because the worksites of construction employers are often difficult to locate, the following means may be used to identify other worksites of the cited employer.

–   If the severe violator enforcement case is resolved through a settlement, the agreement should require the employer to notify the Area Director of its other jobsites prior to when work starts at new construction sites during the following one-year period.

–   An administrative subpoena may be issued to an employer prior to the issuance of the citation to identify the location of worksites where employees of that employer are presently working or are expected to be working within the next 12 months. See *FOM*, Chapter 15, section I, (*Administrative Subpoenas*).

–   A subpoena may be issued at any time during an inspection if it appears that the inspection is likely to result in a SVEP case and the Area Director determines (after consultation with the RA) that the hazards disclosed by the inspection and the inadequacy of the employer's response to those hazards indicate that a broader response by OSHA is appropriate.

– Whenever a subpoena is to be issued pursuant to the SVEP, the Regional Administrator shall coordinate with the Regional Solicitor.

b. National Office.

– When a Regional Administrator determines that a SVEP construction employer is operating in a different region, the Regional Administrator will send a recommendation for inspection, including all relevant facts, to the Director of DEP. The Director shall consult with the Associate Solicitor as appropriate.

– When the Director of DEP deems it necessary to notify Regional Administrators and State Designees regarding activity of a particular construction employer with worksites in more than one Region and/or State Plan States, the Director will issue a SVEP Nationwide referral. The procedures outlined under XV.B.4.b. (p. 13) will be followed.

– Any inspection conducted under a SVEP Nationwide referral is to be coded as an unprogrammed-referral, and is to be considered a referral from the National Office. An OSHA-90 is to be generated when a site is discovered where a SVEP Nationwide Referral employer is operating.

6. Scope of Related Inspections.
The scope of inspection of a related establishment will depend upon the evidence gathered in the original SVEP inspection, and will mainly focus on the same or similar hazards to those found in the original case.

7. Priority of the Inspection.
In accordance with inspection priorities of the FOM (CPL 02-00-148), in Chapter 2, in section IV.B., (*Inspection Priority Criteria*), the SVEP nationwide referral inspections will come after imminent danger, fatality, and complaints, but before other programmed inspections. But see section XVII. (*Relationship to Other Programs*) of this Instruction (p. 19), regarding when other inspections may be conducted concurrently.

The Assistant Secretary by this Instruction determines that the specific inspections under the Nationwide Inspections of Related Workplaces/Worksites process are critical inspections for the purpose of 29 CFR §1908.7(b)(2)(iv) and delegates to the Director of the Directorate of Enforcement Programs the authority to determine site selections of those related workplaces/worksites.

C. Increased Company Awareness of OSHA Enforcement.
1. Sending Citations and Notifications of Penalty to Headquarters.
a. For all employers that are the subject of a SVEP case, the Area Director shall mail a copy of the Citations and Notifications of Penalty to the employer's national headquarters if the employer

has more than one fixed establishment. See sample cover letter in Appendix C.

    b. Employee representatives (e.g., unions) shall receive a copy of the Citations and Notifications of Penalty that is mailed to the employer's national headquarters.

2. Issuing News Releases.

    a. Regional News Releases.

The Regional Offices may issue a News Release for all SVEP cases upon issuance of the citations. Regional Administrators have the discretion to determine which SVEP cases will receive a News Release.

    b. Nationwide Referral Inspection News Releases.

In SVEP cases that were a result of a Nationwide Referral, the Regional Office is to issue a News Release at the time the citations are issued. In certain SVEP cases, the National Office may issue a News Release in coordination with the Regional Office.

3. Sending Letters to Corporate Officers or Coordinating Meetings with the Regional or National Office.

In cases where OSHA determines that an establishment's safety and health problems should be addressed at the corporate level, the following actions should be considered:

    a. A letter sent from the Regional Administrator, or the appropriate National Office official, to the company President expressing OSHA's concern with the company's violations. A copy of the citations shall be sent with the letter if the citations and cover letter have not been sent to the company President previously.

    b. A meeting may be held between OSHA, company officials, employees and unions representing affected employees to discuss how the company intends to address safety and health compliance. If the company operates in more than one region, this normally will require National Office coordination.

    c. Employee representatives shall be notified by letter when OSHA determines that the establishment's safety and health problems need to be addressed at the corporate level.

D. Enhanced Settlement Provisions.

The following settlement provisions shall be considered to ensure future compliance both at the cited facility and at other related facilities of the employer:

1. Employers shall hire a qualified safety and health consultant to develop and implement an effective and comprehensive safety and health program or, where appropriate, a program to ensure full compliance with the subpart under which the employer was cited under the SVEP;

NOTE: Employers cannot be required in a settlement agreement to use OSHA's state consultation services; such services are strictly voluntary.

2. Applying the agreement company-wide (See CPL 02-00-090 Guidelines for Administration of Corporate-wide Settlement Agreements. June 3, 1991);

   NOTE: Company-wide Settlement Agreements are to be coordinated with the National Office of the Solicitor.
3. Requiring interim abatement controls if OSHA is convinced that final abatement cannot be accomplished in a short period of time;
4. In construction (and, where appropriate, in general industry), using settlement agreements to obtain from the employer a list of its current jobsites, or future jobsites within a specified time period. The employer should be required to indicate to OSHA the specific protective measure to be used for each current or future jobsite;
5. Requiring the employer for a specified time period to submit to the Area Director its Log of Work-related Injuries and Illnesses on a quarterly basis, and to consent to OSHA conducting an inspection based on the information;
6. Requiring the employer for a specified time period to notify the Area Office of any serious injury or illness requiring medical attention and to consent to an inspection; and
7. Obtaining employer consent to entry of a court enforcement order under Section 11(b) of the Act.

E. Federal Court Enforcement under Section 11(b) of the OSH Act.
   SVEP cases should be strongly considered for section 11(b) orders when it appears that such orders may be needed to assure compliance. An employer's obligation to abate a cited violation arises when there is a final order of the Review Commission affirming the citation. For guidance on drafting citations and settlement agreements that can maximize the deterrent effect of a Section 11(b) order, see *FOM* (CPL 02-00-148) Chapter 15, section XIV.

XVI. SVEP Log.

A. General.
   The National Office will maintain a SVEP Log in which inspections that meet the SVEP criteria, or are SVEP-related inspections (i.e., SVEP follow-ups, or inspections at other worksites of the same employer), are logged as they are reported to the National Office by the Regional SVEP Coordinators.

B. Lining-Out Establishments from the SVEP Log.
   If an establishment has entered into a settlement agreement (informal or formal) in which a citation that qualified the establishment for SVEP designation is deleted, or if there has been an Administrative Law Judge, Review Commission, or court decision that has vacated such a citation, then the entry on the SVEP Log will be lined-out and the IMIS "SVEP" code will be removed from that establishment's Internet Inspection Detail summary. The Area Director must notify the Regional SVEP Coordinator of these changes, who in turn must notify DEP to line-out the inspection from the SVEP Log.

XVII.   Relationship to Other Programs.
  A.   Unprogrammed Inspections.
       If an unprogrammed inspection arises with respect to an establishment
       that is to receive an SVEP-related inspection, the two inspections may
       be conducted either separately or concurrently. This Instruction does not
       affect in any way OSHA's ability to conduct unprogrammed inspections.
  B.   Programmed Inspections.
       Some establishments selected for inspection under the SVEP may also
       fall under one or more other OSHA initiatives such as Site-Specific
       Targeting (SST) or Local Emphasis Programs (LEP). Inspections under
       these programs may be conducted either separately or concurrently
       with inspections under this Instruction.
XVIII.   Recording and Tracking Inspections.
  A.   SVEP Code.
       This applies to all severe violator enforcement cases issued on or after
       the effective date of this Instruction. Once a case is identified as a severe
       violator enforcement case, enter the NEP code "SVEP" from the drop-
       down list in field 25d, for the inspection.
       NOTE: **Only inspections** that meet one of the four criteria for a
       severe violator enforcement case will be coded with the SVEP NEP
       code.
  B.   NEP Codes for High-Emphasis Hazards.
       If the SVEP criterion used is that described in paragraph XII.B., the
       appropriate NEP codes must be entered in field 25d.
  C.   Significant Enforcement Actions and Enhanced Settlement Codes.
       If any inspection in a significant enforcement action qualifies as a severe
       violator case, it is to be coded "SIGCASE" in item 42, for that inspection.

                    EXAMPLE: N   08   SIGCASE

       If a severe violator case receives an enhanced settlement agreement, it
       is to be coded "ENHSA" in item 42.

                    EXAMPLE: N   08   ENHSA

  D.   Other Program Codes.
       Remember to enter all applicable SST, REP, NEP, and LEP program
       codes in Item 25c and 25d when an inspection is conducted and the
       inspection also meets the protocol for other program(s). Also, enter
       all applicable Strategic Management Plan hazard/industry codes in
       Item 25f.
XIX.   Dun & Bradstreet Number.
       If it is available, the Data Universal Numbering System (DUNS) number is
       to be entered in the appropriate field on the Establishment Detail Screen. In
       establishments where ownership has changed, enter the DUNS number for
       the new owner. If the new owner does not have a new DUNS number, enter

the old DUNS number, if known. Since the DUNS number is site-sensitive, the old number will give some useful data. The field on the Establishment Detail Screen can be accessed by pressing F5 in Item 8 to access establishment processing. Once establishment processing is completed, the DUNS number will appear in Item 9b.

XX. End of the Fiscal Year Report.

The Directorate of Enforcement Programs (DEP) will compile an End of the Fiscal Year Report of each Region's SVEP activity, which will be sent to the Assistant Secretary.

APPENDIX A

INFORMATION NEEDED ON EACH SVEP INSPECTION
FOR MONTHLY REPORT TO THE NATIONAL OFFICE

Employer Name   Inspection Number   Regional Office   Area Office
Opening Date   SIC & NAICS codes   # of Employees Controlled

Indicate if inspection is a SVEP, a Follow-up (FU), a Construction-Related (C-R), or a General Industry-Related (GI-R). If inspection is done based on an SVEP Nationwide Expansion Memo the inspection will either be a C-R or a GI-R.

If the inspection is other than a SVEP, give the name and inspection number of the SVEP case to which it is a follow-up or related.

Remember: any FU, C-R, or GI-R inspections can also be a SVEP.

Indicate if construction or non-construction.

What SVEP criteria apply (more than one can apply):

1) Fatality/Catastrophe—One/more W/R/FTA based on a serious violation of **any gravity** related to death or three or more hospitalized
2) Non-Fatality/Catastrophe—Two/more W/R/FTA based on high gravity serious violations related to a High-Emphasis Hazard (excluding Process Safety Management)
3) Non-Fatality/Catastrophe for PSM hazards—Three/more W/R/FTA based on high gravity serious violations
4) Egregious Case

What SVEP actions have been taken (do not report any planned activities)?:

1) Follow-up inspection conducted; or compelling reason not to conduct
2) Additional construction worksite inspected
3) Additional general industry worksite inspected
4) News Releases issued by Regional Office
5) Letter and citation sent to company headquarters by Region or National Office official
6) Meeting with company officials (separate from informal conference)
7) Enhanced settlement provisions used in informal/formal settlements
8) Court enforcement under Sec. 11(b)
   - Case submitted to RSOL
   - Case submitted to N.O. SOL
   - Petition filed with court [State which court & date]
   - Petition granted by court [State which court & date]
   - Other actions [State & give date]

APPENDIX B

CSHO GUIDANCE: CONSIDERATIONS IN DETERMINING COMPANY STRUCTURE AND SAFETY AND HEALTH ORGANIZATION

When determining whether to inspect other worksites of a company that has been designated a severe violator enforcement case, it must first be determined whether compliance problems and issues found during the initial SVEP inspection are localized or are likely to exist at other, similar facilities owned and operated by that employer. If violations at a local workplace appear to be symptomatic of a broader company neglect for employee safety and health, either generally or with respect to conditions cited under the SVEP inspection, the company structure must be investigated to help identify other establishments and conditions similar to those found in the initial inspection. At the request of the Director of Directorate of Enforcement Programs, a Regional Administrator, or a Regional Coordinator, the Office of Statistical Analysis will be contacted to assist in identifying similar or related worksites of the employer.

Extent of Compliance Problems. Are violative conditions a result of a company decision or interpretation concerning a standard or hazardous condition? Have corporate safety personnel addressed the standard or condition? Ask the following types of questions of the plant manager, safety and health personnel and line employees.

- Who made the decision concerning the violative operation, local management or company headquarters? Was the decision meant to apply to other facilities of the employer as well? If the decision was from company headquarters, what is their explanation?
- Is there a written company-wide safety program? If so, does it address this issue? If so, how is the issue addressed?
- Is there a company-wide safety department? If so, who are they and where are they located? How does company headquarters communicate with facilities/worksites? Are establishment/worksite management and safety and health personnel trained by the company?
- Do personnel from company headquarters visit facilities/worksites? Are visits on a regular or irregular basis? What subjects are covered during visits? Are there audits of safety and health conditions? Were the types of violative conditions being cited discussed during corporate visits?
- Are there insurance company or contractor safety and health audit reports that have been ignored? Are headquarters safety and health personnel aware of the reports and the inaction?
- Does the company have facilities or worksites other than the one being inspected that do similar or substantially similar work, use similar processes or equipment, or produce like products? If so, where are they?
- What is the overall company attitude concerning safety and health? Does the establishment or worksite receive good support from company headquarters on safety and health matters?

- Does the company provide appropriate safety and health training to its employees?
- Ask whether the establishment's/worksite's overall condition is better or worse at present compared to past years? If it is worse, ask why? Has new management or ownership stressed production over safety and health? Is the equipment outdated or in very poor condition? Does management allege that stressed financial conditions keep it from addressing safety and health issues?
- Is there an active and adequately funded maintenance department? Have they identified these problems and tried to fix them?
- Has the management person being interviewed worked at or visited other similar facilities or worksites owned by the company? How was this issue being treated there?

Identifying Company Structure. Inquire where other facilities or worksites are located and how they may be linked to the one being inspected? Sometimes establishment/worksite management will not have a clear understanding of the company structure, just an awareness of facts concerning control and influence from the corporate office.

- Is the establishment/worksite, or the company that owns the establishment or uses the worksite, owned by another legal entity (parent company)? If so, what is the name and location? Try to find out whether the inspected establishment/worksite is a "division" or a "subsidiary" of the parent company. (NOTE: A "division" is a wholly-owned part of the same company that may be differently named, e.g., Chevrolet is a division of GM. A "subsidiary" is a company controlled or owned by another company which owns all or a majority of its shares. Try to determine if the parent company has divisions or subsidiaries other than the one that owns or uses the establishment or worksite being inspected. If so, try to get the names and the type of business they are involved in. Sometimes this type of information can be found on a website or in Dun and Bradstreet. Another good source of information is the office of the Secretary of State within the state government.
- Are there other facilities or worksites controlled by these entities that do the same type of work and might have the same kinds of safety and health concerns?
- Are the company entities publicly held (have publicly traded shares) or are they closely held (owned by one or more individuals)?
- What are the names, positions and business addresses of relevant company personnel of whom interviewees are aware? For which entities do the company safety and health personnel work?
- On what kind of safety and health-related issues or subjects do personnel from company headquarters give instructions?
- Are there other companies owned by the same or related persons that do similar work (especially in construction)?

APPENDIX C

SAMPLE LETTER TO COMPANY HEADQUARTERS

Area Office Header

Date

Name of Employer's National Headquarters

Address of Headquarters

Dear _____:

Enclosed you will find a copy of a Citation and Notification of Penalties for violations of the Occupational Safety and Health Act of 1970, which were issued to [establishment name, located in city, state]. This case has been identified as a severe violator enforcement case under the Occupational Safety and Health Administration's (OSHA) Severe Violator Enforcement Program (SVEP).

The violations referred to in this Citation must be abated by the dates listed and the penalties paid, unless they are contested. This Citation and Notification of Penalties is being provided to you for informational purposes so that you are aware of the violations; the original was mailed to [establishment name] on [date]. We encourage you to work with all of your sites to ensure that these violations are corrected.

OSHA is dedicated to saving lives, preventing injuries and illnesses and protecting America's workers. For more information about OSHA programs, please visit our website at www.osha.gov.

Sincerely,

Area Director

Enclosure

Index